T0251758

Magnetic Resonance Imaging

The Basics

Magnetic Resonance Imaging

The Basics

Christakis Constantinides, PhD

CRC Press
Taylor & Francis Group
Boca Raton   London   New York

CRC Press is an imprint of the
Taylor & Francis Group, an **informa** business

CRC Press
Taylor & Francis Group
6000 Broken Sound Parkway NW, Suite 300
Boca Raton, FL 33487-2742

© 2014 by Taylor & Francis Group, LLC
CRC Press is an imprint of Taylor & Francis Group, an Informa business

No claim to original U.S. Government works

Printed on acid-free paper
Version Date: 20131217

International Standard Book Number-13: 978-1-4822-1731-5 (Paperback)

This book contains information obtained from authentic and highly regarded sources. While all reasonable efforts have been made to publish reliable data and information, neither the author[s] nor the publisher can accept any legal responsibility or liability for any errors or omissions that may be made. The publishers wish to make clear that any views or opinions expressed in this book by individual editors, authors or contributors are personal to them and do not necessarily reflect the views/opinions of the publishers. The information or guidance contained in this book is intended for use by medical, scientific or health-care professionals and is provided strictly as a supplement to the medical or other professional's own judgement, their knowledge of the patient's medical history, relevant manufacturer's instructions and the appropriate best practice guidelines. Because of the rapid advances in medical science, any information or advice on dosages, procedures or diagnoses should be independently verified. The reader is strongly urged to consult the drug companies' printed instructions, and their websites, before administering any of the drugs recommended in this book. This book does not indicate whether a particular treatment is appropriate or suitable for a particular individual. Ultimately it is the sole responsibility of the medical professional to make his or her own professional judgements, so as to advise and treat patients appropriately. The authors and publishers have also attempted to trace the copyright holders of all material reproduced in this publication and apologize to copyright holders if permission to publish in this form has not been obtained. If any copyright material has not been acknowledged please write and let us know so we may rectify in any future reprint.

Except as permitted under U.S. Copyright Law, no part of this book may be reprinted, reproduced, transmitted, or utilized in any form by any electronic, mechanical, or other means, now known or hereafter invented, including photocopying, microfilming, and recording, or in any information storage or retrieval system, without written permission from the publishers.

For permission to photocopy or use material electronically from this work, please access www.copyright.com (http://www.copyright.com/) or contact the Copyright Clearance Center, Inc. (CCC), 222 Rosewood Drive, Danvers, MA 01923, 978-750-8400. CCC is a not-for-profit organization that provides licenses and registration for a variety of users. For organizations that have been granted a photocopy license by the CCC, a separate system of payment has been arranged.

Trademark Notice: Product or corporate names may be trademarks or registered trademarks, and are used only for identification and explanation without intent to infringe.

Library of Congress Cataloging-in-Publication Data

Constantinides, Christakis, author.
  Magnetic resonance imaging : the basics / Christakis Constantinides.
  p. ; cm.
  Includes bibliographical references and index.
  ISBN 978-1-4822-1731-5 (paperback : alk. paper)
  I. Title.
  [DNLM: 1. Magnetic Resonance Imaging. 2. Magnetic Resonance Spectroscopy. WN 185]

  RC386.6.M34
  616.07'548--dc23   2013048808

Visit the Taylor & Francis Web site at
http://www.taylorandfrancis.com

and the CRC Press Web site at
http://www.crcpress.com

I dedicate this book to my mother, Elpida, and grandfather, Christodoulos Tapakoudes, who have taught me the basic principles of human respect, dignity, and humility.

Contents

5 Fundamentals of Magnetic Resonance III: The Formalism of k-Space   53

6 Pulse Sequences   61

7 Introduction to Instrumentation   75

Foreword

Who could have predicted the impact of Otto Stern's discovery that the proton possesses a magnetic moment? The physics community was sufficiently impressed that Professor Stern was awarded the 1943 Nobel Prize in Physics. A cascade of Nobel Prizes followed—first in physics (Isidor Isaac Rabi, 1944; Felix Bloch and E.M. Purcell, 1952), then in chemistry, as the fundamental principles were extended to understand molecular structure (Richard Ernst, 1991; Kurt Wüthrich, 2002), and finally to medicine (Paul Lauterbur and Peter Mansfield, 2003). And there is every reason to expect further Nobel awards as the creative minds of the next generation of scientists find new ways to exploit this fundamental phenomenon.

This book, by Professor Constantinides, distills the work of so many of these giants and provides the foundation for a new generation of Nobel laureates. It brings together in one text a wonderfully coherent story of the history of nuclear magnetic resonance (NMR), the physics of magnetic resonance, the mathematics of the Fourier transform employed for spatial encoding and image reconstruction, the chemistry of NMR spectroscopy, the biology that can be revealed with in vivo spectroscopy, and the engineering principles underlying system design—all with abundant clinical examples. It truly does provide "everything you need to know." The book covers these topics with both breadth and depth. The engineer, for whom the Fourier transform is clear, can appreciate the origins of contrast through a discussion of relaxation and the molecular environment. The biochemist probing the energetics of phosphorous can appreciate the potential limitations in his or her data from undersampling during spatial encoding. The neuroscientist exploiting functional magnetic resonance imaging (fMRI) to understand that complexity of human thought must appreciate the limitations to the BOLD signal derived from changes in blood oxygenation. And a physician reaching out to understand molecular changes in the clinical setting can appreciate the limits and opportunities for new engineering approaches.

The text is perfectly compartmentalized and ordered. The new entrant to the field can start at the beginning and work his or her way through in a structured fashion, with each chapter building on the previous one. The organization of the chapters allows the scientist with some background to easily jump into the middle. For example, the chapter on image encoding (Chapter 4) provides a compact discussion of spatial resolution for the chemist who already understands relaxation. The book provides an excellent map for a graduate engineering course, with many chapters suitable for undergraduates. Many chapters are well suited for residents and clinicians who must face the mysteries of MRI on a daily basis. Above all, this text is integrative and brings together in a single, compact, and well-written format everything you need to know about MRI.

G. Allan Johnson, PhD
Charles E. Putman University Professor of Radiology, Physics, and
Biomedical Engineering
Director, Center for In Vivo Microscopy
Duke University Medical Center

Book Synopsis

Magnetic resonance imaging (MRI) is a rapidly developing field in basic applied science and clinical practice. Research efforts in this field have already been recognized with five Nobel prizes, awarded to seven Nobel laureates during the last 70 years. The book begins with a general description of the phenomenon of magnetic resonance and a brief summary of Fourier transformations in two dimensions. It proceeds to examine the fundamental principles of physics for nuclear magnetic resonance (NMR) signal formation and image construction. To this extent, there is a detailed reference to the mathematical formulation of MRI using the imaging equation, description of the relaxation parameters T_1 and T_2, and reference to specific pulse sequences and data acquisition schemes. Additionally, numerous image quantitative indices are presented, including signal, noise, signal-to-noise, contrast, and resolution. The second part of the book discusses the hardware and electronics of an MRI scanner, the typical measurements and simulations of magnetic fields based on the law of Biot–Savart, followed by an introduction to NMR spectroscopy, and to dedicated spectral techniques employing various pulse sequences. The third part discusses advanced imaging techniques. While the list may contain numerous modern applications, including cardiac MR, coronary and peripheral angiography, flow, diffusion, and functional MRI (fMRI), the focus is maintained on parallel imaging. The book is enriched with numerous worked examples and problem sets with selected solutions.

Nobel Prizes in Magnetic Resonance

Magnetic resonance imaging is a field that emerged right after the Second World War, as a result of experimental work that was initiated for spectroscopy. It progressed and flourished rapidly to find its way into clinical diagnosis and practice. It has had tremendous impact in medicine as one of the few noninvasive, nonionizing modalities with excellent tissue contrast and resolution, and for its applicability in cerebrovascular disease. Research and technological efforts and breakthroughs in this field have been numerous over the last 70 years, since the discovery of the phenomenon, and such efforts have been honored, thus far, with five Nobel Prizes awarded to seven laureates.

A recent announcement from the International Society of Magnetic Resonance in Medicine (ISMRM) summarizes the evolution of the major scientific developments and advancements in the field as follows.

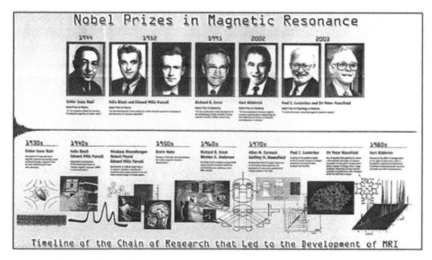

Reproduced from the International Society for Magnetic Resonance in Medicine (ISMRM) poster publication with permission.

Isidor Isaac Rabi—Nobel Prize in Physics, 1944

1930s: Development of molecular beam magnetic resonance by passing a beam of lithium through a magnetic field and then bombarding the beam with radiowaves.

Felix Bloch and Edward Mills Purcell—Nobel Prize in Physics, 1952

1940s: Independent demonstration of the phenomenon known as nuclear magnetic resonance (NMR) in condensed matter.

Richard R. Ernst —Nobel Prize in Chemistry, 1991

1960s: Proof that Fourier analysis of pulsed NMR signals provides increased sensitivity and flexibility over continuous wave NMR methods.

Kurt Wüthrich—Nobel Prize in Chemistry, 2002

1980s: Discovery of the effect of deoxygenation on the signal of select amino acids in hemoglobin. Establishment of methods for determining the structure of proteins and other macromolecules.

Paul Lauterbur and Sir Peter Mansfield—Nobel Prize in Medicine and Physiology, 2003

1990s: Coupling of the gradient concept to the CT scanner of multiple projections and reconstruction to obtain the first MRI (Lauterbur). Use of magnetic field gradients to obtain a 1D projection from layers of camphor. Conception of echo planar imaging that yielded complete 2D images from a single excitation. Origination of slice-selective excitation and publication of the first MRI from a human (Mansfield).

Introduction

This book is designed for undergraduate, graduate, and postgraduate students, as well as medical students, residents, and trainees who seek an in-depth knowledge of the basics of magnetic resonance imaging. It is intended to bridge the gap between excellent classic textbooks in the field, adopting a more simplistic approach. Its focus is on the principles and physics of nuclear magnetic resonance, imaging processing, reconstruction, and hardware systems and instrumentation. It requires prior knowledge in simple and advanced mathematics, linear systems, and image processing, as well as a basic knowledge of electrical circuits. Toward the end, the book attempts to introduce biomedical engineers, physicists, or radiologists to parallel imaging, an advanced imaging and novel MRI technique.

The book's structure and content are based on excellent sources and prior courses on medical imaging and magnetic resonance taught at the Johns Hopkins University (Professors E. McVeigh, P. Bottomley, and P. Barker, to whom the author was instrumental while serving as a teaching assistant in 1996) and categorical courses offered at the scientific meetings of the International Society of Magnetic Resonance in Medicine (ISMRM). The book is enriched with multiple worked examples and numerous problem sets contributed by the author and Professor E. McVeigh, and Drs. S. Reeder and B. Bolster (fellow graduate students at Johns Hopkins during 1992–2000), for which the author is most grateful. It is hoped that this textbook will serve as a reference guide to magnetic resonance imaging teaching and learning, and that it will complement reading, in addition to the plethora of other excellent textbooks in the field.

About the Author

Christakis Constantinides, PhD, completed his undergraduate studies at the Imperial College of Science, Technology, and Medicine (bachelor's with first-class honors in electrical and electronic engineering, 1992) and his graduate (CASP/USIA scholar) and postgraduate (Whitaker Foundation scholar) studies at the Johns Hopkins University Whiting School of Engineering (MSE, 1994) and the Johns Hopkins University School of Medicine (PhD, 2000) in biomedical engineering. He held an appointment as a visiting research fellow (Fogarty International Fellow) at the National Institutes of Health (2001–2003). He joined the faculty of the Mechanical Engineering Department at the University of Cyprus in September 2005. His specific research interests focus on the study of cardiac mechanical function, computational and tissue structure modeling and characterization, hardware design, and functional and cellular tracking methods using magnetic resonance imaging. The goal of his research efforts is the complete characterization of the electromechanical function of the heart in small animals and humans, aiming to promote the understanding of mechanisms of human disease that is predominantly underlined by genetic causes. Dr. Constantinides has taken an appointment as an assistant professor in the Department of Mechanical Engineering and has acted as the director of the ISO 9001 certified Laboratory of Physiology and Biomedical Imaging (LBI) "HIPPOCRATES." He teaches in the areas of mechatronics and automated systems, and in biomedical physiology and imaging.

Dr. Constantinides has published over 60 papers in international peer-reviewed journals and conferences. He is a member of the American Physiological Society (APS), the International Society of Magnetic Resonance in Medicine (ISMRM), and serves as a reviewer for *Magnetic Resonance in Medicine*, *Journal of Magnetic Resonance*, *NMR in Biomedicine*, and *Journal of Applied Physiology*.

List of Abbreviations

1D	One-dimensional
2D	Two-dimensional
3D	Three-dimensional
AM	Amplitude modulation
BIBO	Bounded input–bounded output
CNR	Contrast-to-noise ratio
CPMG	Carr–Purcell–Meiboom–Gill
CSI	Chemical shift imaging
CW	Continuous wave
DC	Direct current
DFT	Discrete Fourier transform
DICOM	Digital imaging and communications in medicine
DRESS	Depth-resolved surface coil spectroscopy
ECG	Electrocardiogram
EMF	Electromotive force
EPI	Echo planar imaging
FDA	Food and Drug Administration
FID	Free induction decay

FLASH	Fast low-angle shot
fMRI	Functional MRI
FOV	Field of view
FSE	Fast spin–echo
FT	Fourier transform
FWHM	Full width at half maximum
GE	General Electric
GRAPPA	Generalized autocalibrating partially parallel acquisition
GRASS	Gradient recalled at steady state
GRE	Gradient echo
IEC	International Electrochemical Company
IRB	Institutional review board
ISIS	Image selected in vivo spectroscopy
ISMRM	International Society of Magnetic Resonance in Medicine
LSI	Linear shift invariant
MHz	Megahertz
MRE	Magnetic resonance elastography
MRI	Magnetic resonance imaging
NMR	Nuclear magnetic resonance
NOE	Nuclear Overhauser effect
PCr	Phosphocreatine
PILS	Partially parallel imaging with localized sensitivities
PPM	Parts per million
PRESS	Point-resolved surface coil spectroscopy
PSF	Point spread function
PSS	Point spread sequence
RF	Radio frequency
RFZ	Rotating-frame zeygmatography

SAR	Specific absorption rate
SENSE	Sensitivity encoding
SMASH	Sensitivity acquisition at spatial harmonics
SNR	Signal-to-noise ratio
SPACE RIP	Sensitivity profiles from an array of coils for encoding and reconstruction in parallel
SSFP	Steady-state free precession
STEAM	Stimulated echo acquisition mode
TE	Echo time
TI	Inversion time
TM	Mixing time
TMS	Trimethylsilane
TR	Repetition time

List of Symbols

A	Anterior
Å	Angstrom (10^{-10} m)
α	Rotation angle
B_o	Static magnetic field
BW	Bandwidth
B_1	The radio frequency magnetic field
B_{1R}	Receive B_1 magnetic field
B_{1T}	Transmit B_1 magnetic field
C	Contrast
CW	Continuous wave
dB	Decibel
DBM	Doubly balanced mixer
E	Energy
EPI	Echo planar imaging
ESR	Electron spin resonance
FID	Free induction decay
fMRI	Functional MRI
f_o	Resonance frequency in Hertz
FT	Fourier transform
G_{BP}	Bipolar magnetic field gradient
G_f	Frequency encoding gradient
G_i	Field gradient in the i direction
G_s	Slice selection gradient
G_y	Phase encoding gradient
$G_{y,\,max}$	Maximum value of phase encoding gradient
γ	Gyromagnetic ratio
h	Planck's constant
I	Inferior
IFT	Inverse Fourier transform

IM	Imaginary part of a complex number
J	Joule
k	Boltzmann constant; proportionality constant
k	Kilo (10^3)
K	Kelvin temperature
L	Left
LUT	Lookup table
m	Milli (10^{-3})
M_o	Equilibrium magnetization
M_X	X component of magnetization
$M_{X'}$	X' component of magnetization
M_Y	Y component of magnetization
$M_{Y'}$	Y' component of magnetization
M_{XY}	Transverse component of magnetization
M_Z	Z component of magnetization
μ	Micro (10^{-6})
MRA	Magnetic resonance angiography
MRI	Magnetic resonance imaging
N^+	Spin population in low energy state
N^-	Spin population in high energy state
NEX	Number of excitations (number of averages)
NMR	Nuclear magnetic resonance
ω_o	Resonance frequency in radians per second (Larmor frequency)
P	Posterior; power
ppm	Parts per million
φ	Phase angle
π	3.14159 …
Q	RF coil quality factor
R	Right
RE	Real part of a complex number
RF	Radio frequency
ρ	Proton density
S	Superior
s	Second
SAR	Specific absorption rate
Sinc	$Sin(x)/x$
SNR	Signal-to-noise ratio
STS	Single-turn solenoid
T	Temperature
T	Tesla
T_1	Spin–lattice relaxation time
T_2	Spin–spin relaxation time
T_2^*	T_2 star
$T_{2,\,inhomo}$	Inhomogeneous T_2
Thk	Slice thickness

X	Axis in laboratory coordinate system
X'	Rotating frame X axis
Y	Axis in laboratory coordinate system
Y'	Rotating frame Y axis
Z	Axis in laboratory coordinate system

1 Fourier Transformations

1.1 Introduction

Fourier transformation is a powerful mathematical operation. It was named after Jean Baptiste Joseph Fourier, a French mathematician and physicist (1768–1830) who initiated the investigation of Fourier series and its application to problems of heat flow. This integral transform reexpresses a function in terms of its sinusoidal basis functions as a scaled summation. The advent of electronics and the evolution of computers in the early 20th century allowed computation and digital signal processing using the Fourier transform (FT) in the discrete domain.

Interestingly, in the 1960s, Professor Richard Ernst (35; Ernst 1966, 34; Ernst 1987) at the ETH in Zurich applied the basic Fourier transformation in its discrete form to formulate the magnetic resonance imaging (MRI) signal evolution in pulse-acquire experiments. His brilliant work led to the proof that Fourier analysis of pulsed nuclear magnetic resonance (NMR) signals provides increased sensitivity over continuous wave NMR methods. For his groundbreaking work, which revolutionized the way modern scanner systems acquire data and reconstruct them, he was awarded the Nobel Prize in Chemistry in 1991.

This chapter introduces basic concepts of the continuous and discrete Fourier transform and its application to two-dimensional images.

1.2 Mathematical Representation of Images

An image $I(x, y)$ is a two-dimensional signal represented by a two-dimensional array of the independent variables x and y. A pixel (picture element) is the smallest element representation of the two-dimensional array in an image.

Images can be classified in two categories, continuous and discrete. A continuous image $I(x, y)$ is a two-dimensional function of variables x and y, which take values within a continuous range in the two-dimensional space R^2.

A discrete image $I(n, m)$ is a function of two independent variables n and m, which are only defined at specific points in the two-dimensional space.

Images are typically presented in gray scale, with shades of gray mapped to the value of the continuous or discrete variables x, y or n, m. Some diagnostic modalities (such as positron emission tomography (PET) or single-photon emission computed tomography (SPECT)) use color to represent (or map) the signal intensities of the pixels for representation and display. Medical images are usually binary images, with each pixel represented by a number of gray-level values. A smaller gray-level value typically represents a darker gray shade.

Digital Imaging and Communications in Medicine (DICOM) is the universal standard image format for saving medical diagnostic images. So, for example, for a 16-bit DICOM image, 2^{16} levels of gray are needed to represent the entire dynamic range of image pixel intensities.

1.3   Continuous Images

A two-dimensional image $I(x, y)$ is a mathematical function of the independent variables x and y such that

$$I(x, y), \quad -\infty < x, y < +\infty \tag{1.1}$$

1.4   Delta Function

Following the definition of a one-dimensional delta (or Fermi–Dirac or Kronecker-delta) function $\delta(x)$, or $\delta(y)$, a two-dimensional delta function $\delta(x, y)$ is defined such that (Macovski 1983; Jain 1989)

$$\int_{-\infty}^{+\infty} \int_{-\infty}^{+\infty} \delta(x, y)\, dx\, dy = 1 \tag{1.2}$$

1.5   Separable Images

An image $I(x, y)$ is defined to be separable if and only if (iff) there exist two functions $I(x)$ and $I(y)$ such that

$$I(x, y) = I(x).I(y) \qquad\qquad -\infty < x, y < +\infty \tag{1.3}$$

The two-dimensional delta function $\delta(x, y)$ is separable iff

$$\delta(x, y) = \delta(x).\delta(y) \tag{1.4}$$

1.6   Linear Shift Invariant (LSI) Systems

Using a black box representation of a system, for every input there is a corresponding output characteristic to its response function R. The response function uniquely characterizes the system such that

$$G(x, y) = R[I(x, y)] \tag{1.5}$$

A system is defined to be linear if the response output to a linearly scaled input is the scaled summation of the responses of the constitutive parts:

$$R[\alpha.I_1(x, y) + \beta.I_2(x, y)] = \alpha.R[I_1(x, y)] + \beta.R[I_2(x, y)] = \alpha G_1(x, y) + \beta.G_2(x, y) \quad (1.6)$$

A linear system is uniquely characterized by its *impulse response* or *point spread function* (PSF), $h(x, y)$, that is, the system response to a unit impulse (or delta function) input, in accordance to (Goutsias 1996)

$$h(x, y; \chi, \psi) = R[\delta(x - \chi, y - \psi)] \quad (1.7)$$

For a linear system,

$$J(x,y) = \int_{-\infty}^{+\infty}\int_{-\infty}^{+\infty} h(x, y; \chi, \psi)I(\chi, \psi)\,d\chi\,d\psi \quad (1.8)$$

$J(x, y)$ is known as the *superposition integral*. Equivalently, it can be stated that a system's frequency response is the Fourier transform of its impulse response.

Correspondingly, a system is known as *shift invariant* if

$$G(x - x_o, y - y_o) = R[I(x - x_o, y - y_o)] \quad (1.9)$$

for every pair (x_o, y_o). In simpler words, a shifted input will result in a shifted output response. Equivalently, for a linear shift invariant (LSI) system, with the input being the impulse response,

$$G(x,y) = \int_{-\infty}^{+\infty}\int_{-\infty}^{+\infty} h(x - \chi, y - \psi)I(\chi, \psi)\,d\chi\,d\psi \quad (1.10)$$

Alternatively:

$$G(x, y) = h(x, y)**I(x, y) \quad (1.11)$$

also known as the convolution integral.

1.7 Cascade Systems

Linear and nonlinear systems are often cascaded, either in series or in parallel.

1.7.1 Serial Cascade of LSI Systems

If $h_1(x,y)$ and $h_2(x,y)$ are the impulse responses of two LSI systems cascaded serially (Figure 1.1), it can then be written that (Goutsias 1996)

$$G(x, y) = h_2(x, y)**[h_1(x, y)**I(x, y)] \quad (1.12)$$

$$G(x, y) = h_1(x, y)**[h_2(x, y)**I(x, y)] \quad (1.13)$$

$$G(x, y) = [h_1(x, y)**h_2(x, y)]**I(x, y) \quad (1.14)$$

$$G(x, y) = h_T(x, y)**I(x, y) \quad (1.15)$$

where

$$h_2(x, y)**h_1(x, y) = h_1(x, y)**h_2(x, y) = h_T(x, y) \quad (1.16)$$

based on well-known properties of the convolution integral.

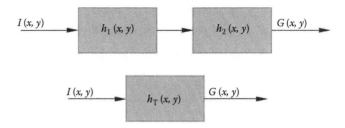

Figure 1.1. Impulse response function of serially cascaded linear shift invariant systems.

Example 1.1 Prove that the total response of serially cascaded LSI systems is the convolution of each of the system responses.

ANSWER

Let

$$N = I ** h_1 \qquad (Ex\ 1.1)$$

$$G = N ** h_2 = I ** h_1 ** h_2 = I ** (h_1 ** h_2) = h_T ** I \qquad (Ex\ 1.2)$$

where

$$h_T = h_1 ** h_2$$

$$N = \int_{-\infty}^{\infty} \int_{-\infty}^{\infty} h_1(x-\kappa, y-\lambda) I(\kappa, \lambda)\, d\kappa\, d\lambda \qquad (Ex\ 1.3)$$

$$G = \int_{-\infty}^{\infty} \int_{-\infty}^{\infty} h_1(x-X, y-Y) N(X, Y)\, dX\, dY \qquad (Ex\ 1.4)$$

$$= \int_{-\infty}^{\infty} \int_{-\infty}^{\infty} h_1(x-X, y-Y)\left[\int_{-\infty}^{\infty} \int_{-\infty}^{\infty} h_1(x-\kappa, y-\lambda) I(\kappa, \lambda)\, d\kappa\, d\lambda \right] dX\, dY \qquad (Ex\ 1.5)$$

$$= \int_{-\infty}^{\infty} \int_{-\infty}^{\infty} \int_{-\infty}^{\infty} \int_{-\infty}^{\infty} \left[h_1(x-\kappa, y-\lambda) h_2(x-X, y-Y)\, d\kappa\, d\lambda \right] I(\kappa, \lambda)\, dX\, dY \qquad (Ex\ 1.6)$$

Let $x - X = \kappa$ and $y - Y = \lambda$

$$G = \int_{-\infty}^{\infty} \int_{-\infty}^{\infty} \left[\int_{-\infty}^{\infty} \int_{-\infty}^{\infty} h_1(x-\kappa, y-\lambda) h_2(\kappa, \lambda)\, d\kappa\, d\lambda \right] I(\kappa, \lambda)\, dX\, dY \qquad (Ex\ 1.7)$$

$$= \int_{-\infty}^{\infty} \int_{-\infty}^{\infty} [h_1 ** h_2] I(\kappa, \lambda)\, dX\, dY = \int_{-\infty}^{\infty} \int_{-\infty}^{\infty} h_T(X, Y) I(x-X, y-Y)\, dX\, dY = h_T ** I \qquad (Ex\ 1.8)$$

1.7.2 Parallel Cascade of LSI Systems

If $h_1(x,y)$ and $h_2(x,y)$ are the impulse responses of two LSI systems cascaded in parallel (Figure 1.2), it can be written that

$$G(x, y) = h_1(x, y)**I(x, y) + h_2(x, y)**I(x, y) \tag{1.17}$$

$$G(x, y) = [h_1(x, y) + h_2(x, y)]**I(x, y) \tag{1.18}$$

$$G(x, y) = h_T(x, y)**I(x, y) \tag{1.19}$$

where

$$h_T(x, y) = h_1(x, y) + h_2(x, y) \tag{1.20}$$

based on the properties of the convolution integral.

Example 1.2 Prove that the total response of parallel cascaded LSI systems is the summation of each of the responses.

ANSWER

Let

$$G = I **h_1 + I **h_2 \tag{Ex 1.9}$$

$$G = \int_{-\infty}^{\infty}\int_{-\infty}^{\infty} h_1(x-X, y-Y)I(X,Y)\,dXdY + \int_{-\infty}^{\infty}\int_{-\infty}^{\infty} h_2(x-X, y-Y)I(X,Y)\,dXdY \tag{Ex1.10}$$

$$= \int_{-\infty}^{\infty}\int_{-\infty}^{\infty} [h_1(x-X, y-Y) + h_2(x-X, y-Y)]I(X,Y)\,dXdY \tag{Ex 1.11}$$

$$= \int_{-\infty}^{\infty}\int_{-\infty}^{\infty} h_T(x-X, y-Y)I(X,Y)\,dXdY \tag{Ex 1.12}$$

$$= h_T **I(X, Y) \tag{Ex 1.13}$$

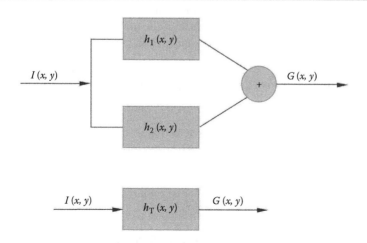

Figure 1.2. Impulse response function of systems in parallel cascade arrangement.

1.8   Stability

An LSI system is defined to be a *bounded input bounded output* (BIBO) stable system if the PSF is absolutely summable, that is:

$$\int_{-\infty}^{+\infty}\int_{-\infty}^{+\infty}|h(x,y)|\,dxdy \quad < +\infty \tag{1.21}$$

1.9   Fourier Transformation and Inverse FT

For a two-dimensional image $i(x, y)$, the Fourier transformation is the mathematical operation defined by

$$FT\big[i(x,y)\big]=I(\omega_x,\omega_y)=\int_{-\infty}^{+\infty}\int_{-\infty}^{+\infty}I(x,y)e^{-j(\omega_x x+\omega_y y)}\,dxdy \tag{1.22}$$

for $-\infty < \omega_x, \omega_y < +\infty$ where $j=\sqrt{-1}$.

In imaging ω_x, ω_y are spatial cyclic frequencies in the horizontal and vertical dimensions (in rad/mm), defined by

$$\omega_x = 2\pi f_x \tag{1.23}$$

and

$$\omega_y = 2\pi f_y \tag{1.24}$$

Frequencies f_x and f_y are known as the horizontal and vertical spatial frequencies (with units in 1/mm).

In a similar fashion, the inverse Fourier transform in defined as

$$FT^{-1}[I(\omega_x,\omega_y)]=\int_{-\infty}^{+\infty}\int_{-\infty}^{+\infty}I(\omega_x,\omega_y)e^{j(\omega_x x+\omega_y y)}\,d\omega_x\,d\omega_y \tag{1.25}$$

Examples of two-dimensional images and the corresponding 2D-FT function are shown in Figure 1.3.

1.10   Properties of Fourier Transformations

In computing Fourier transformations the numerous properties of this mathematical transformation can be invoked. Some of the most interesting and practically useful transformations (Table 1.1) include the following.

Example 1.3 Prove the linearity property of FT:

$$FT\big[\alpha i_1(x,y)+\beta i_2(x,y)\big]=\int_{-\infty}^{\infty}\int_{-\infty}^{\infty}\big[\alpha i_1(x,y)+\beta i_2(x,y)\big]e^{-i(\omega_x x+\omega_y y)}\,dxdy \tag{Ex 1.14}$$

$$=\alpha\int_{-\infty}^{\infty}\int_{-\infty}^{\infty}i_1(x,y)e^{-i(\omega_x x+\omega_y y)}\,dxdy+\beta\int_{-\infty}^{\infty}\int_{-\infty}^{\infty}i_2(x,y)e^{-i(\omega_x x+\omega_y y)}\,dxdy \tag{Ex 1.15}$$

$$=\alpha I_1(\omega_x,\omega_y)+\beta I_2(\omega_x,\omega_y) \tag{Ex 1.16}$$

Figure 1.3. Two-dimensional images and their Fourier transforms. (Reproduced from Professor J. Brayer, University of New Mexico. With permission.)

Table 1.1. Summary Listing of the Most Important Properties of Continuous Fourier Transformations

Property	Transformation
Linearity	$FT[\alpha.i_1(x, y) + \beta.i_2(x, y)] = \alpha FT[I_1(x, y)] + \beta FT[I_2(x, y)] = \alpha I_1(\omega_x, \omega_y) + \beta I_2(\omega_x, \omega_y)$
Scaling	$FT[i(\alpha x, \beta y)] = \dfrac{1}{|\alpha\beta|} I\left(\dfrac{\omega_x}{\alpha}, \dfrac{\omega_y}{\beta}\right)$
Shifting	$FT[i(x - \alpha, y - \beta)] = I(\omega_x, \omega_y).e^{-j(\omega_x\alpha+\omega_y\beta)}$
Convolution	$FT[i_1(x, y)**i_2(x, y)] = I_1(\omega_x, \omega_y).I_2(\omega_x, \omega_y)$
Correlation	$FT[\int_{-\infty}^{+\infty}\int_{-\infty}^{+\infty} i_1(\chi,\psi)i_2^*(x+\chi,y+\psi)d\chi\,d\psi] = I_1(\omega_x,\omega_y)I_2^*(\omega_x,\omega_y)$
Separability	$FT[i_1(x, y).i_2(x, y)] = I_1(\omega_x). I_1(\omega_y). I_2(\omega_x). I_1(\omega_y)$
Parseval's theorem	$\int_{-\infty}^{+\infty}\int_{-\infty}^{+\infty} |i(x,y)|^2 dx\,dy = \dfrac{1}{4\pi^2}\int_{-\infty}^{+\infty}\int_{-\infty}^{+\infty} |I(\omega_x,\omega_y)|^2 d\omega_x\,d\omega_y$

1.11 Frequency Response

The Fourier transform of the PSF of an LSI system, $h(x, y)$ is known as its frequency response, such that

$$H(\omega_x,\omega_y) = \int_{-\infty}^{+\infty}\int_{-\infty}^{+\infty} h(x,y).e^{-j(\omega_x x+\omega_y y)}\,dx\,dy \qquad (1.26)$$

$$h(x,y) = \int_{-\infty}^{+\infty} \int_{-\infty}^{+\infty} H(\omega_x, \omega_y) e^{j(\omega_x x + \omega_y y)} \, d\omega_x \, d\omega_y \qquad (1.27)$$

The frequency response (and in general, operations and analysis in the frequency domain) is critical to the analysis of the behavior of linear systems.

1.12   Discrete Images and Systems

Processing of images is achieved by discretization of the continuous images. To discretize a 2D image, it is sampled in 2D space, most often using conventional rectilinear sampling schemes.

The 2D output image $i(m, n)$ is mathematically defined by

$$i(m, n) = i(m\Delta x, n\Delta y), \text{ for } m, n = ...,-3, -2, -1, 0, 1, 2, 3... \qquad (1.28)$$

where Δx, Δy are known as the horizontal and vertical sampling periods.

The sampling scheme depicted in Figure 1.4 is also known as lexicographic ordering or sampling (Goutsias 1996). The outcome of such a sampling process is the two-dimensional array $i(m, n)$ of discrete numbers that contain the values of the continuous 2D image $i(x,y)$, at the discrete sampling intervals.

1.13   Separable Images

Similar to the definition listed in Section 1.5 for separable continuous images, separability for discrete images can be defined such that

$$i(m, n) = i(m) \, i(n), \text{ for every } m, n \qquad (1.29)$$

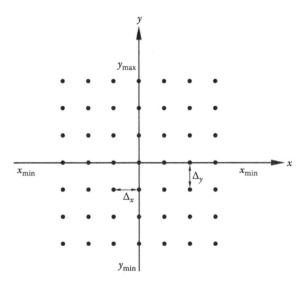

Figure 1.4. Sampling scheme of a continuous 2D image $i(x, y)$ in two dimensions along a rectilinear grid at uniform spatial intervals. Some unconventional sampling schemes and k-space sampling trajectories are introduced in the k-space formalism section.

An image with a finite extent is an image that is bound spatially, that is, one that is zero outside a region of finite extent. This region is known in general as the region of support. In magnetic resonance, images are spatially bound. These bounds define the image field of view (FOV). Such a region of support (or equivalently FOV) can (and often is) asymmetric in the two spatial dimensions.

1.14   Linear Shift Invariant Systems

In a similar fashion to the continuous domain case, a two-dimensional discrete system transforms an input discrete image $i(m, n)$ into an output image $g(m, n)$ according to the system's response function D.

A two-dimensional discrete LSI system is defined accordingly as

$$D[\alpha.i_1(m, n) + \beta.i_2(m, n)] = \alpha.D[i_1(m, n) + \beta.D[i_2(m, n)] = \alpha.g_1(m, n) + \beta.g_2(m, n)$$

(1.30)

and

$$D[i(m - m_0, n - n_0)] = g(m - m_0, n - n_0) \tag{1.31}$$

for every input image $i_1(m, n)$, $i_2(m, n)$ and discrete integers m_0, n_0 such that

$$D[i(m, n)] = g(m, n) \tag{1.32}$$

1.15   Frequency Response: Point Spread Sequence

The point spread sequence (PSS) for an LSI system, $h(m, n)$, is defined according to the convolution summation:

$$g(m,n) = \sum_{m_0}^{+\infty} \sum_{n_0}^{+\infty} h(m - m_0, n - n_0) i(m_o, n_o) \tag{1.33}$$

or equivalently:

$$g(m, n) = h(m, n)**i(m, n) \tag{1.34}$$

The frequency response is simply the discrete Fourier transform of the PSS.

1.16   Discrete Fourier Transform and Its Inverse

The discrete Fourier transform must thus be defined as

$$FT[(i(m,n)] = I(\omega_x, \omega_y) = \sum_{m=-\infty}^{+\infty} \sum_{n=-\infty}^{+\infty} i(m,n).e^{-j(\omega_x m + \omega_y n)} \quad \text{for} \quad -\pi \le \omega_x, \omega_y \le \pi \tag{1.35}$$

Correspondingly, the inverse discrete Fourier transform is given by

$$FT^{-1}[I(\omega_x, \omega_y)] = i(m,n) = \frac{1}{4\pi^2} \int_{-\pi}^{\pi} \int_{-\pi}^{\pi} I(\omega_x, \omega_y).e^{j(\omega_x m + \omega_y n)} \, d\omega_x \, d\omega_y \tag{1.36}$$

Similar to the continuous case,

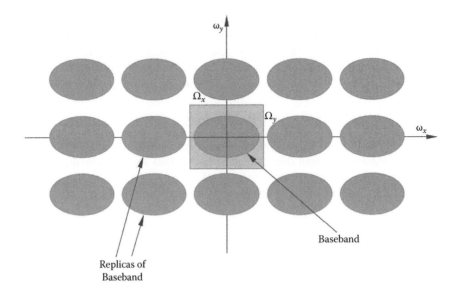

Figure 1.5. Fourier transform pattern (baseband and replicas of finite extent (Ω_x, Ω_y) frequency content of image) of a sampled image with a sampling rate higher than the Nyquist rate. Reconstructing the original image involves low-pass filtering of the baseband.

$$\omega_x = 2\pi f_x \qquad (1.37)$$

$$\omega_y = 2\pi f_y \qquad (1.38)$$

where ω_x, ω_y, f_x, and f_y, represent the horizontal and vertical discrete cyclic and spatial frequencies, respectively. The Fourier transform is periodic with horizontal and vertical sampling periods being equal to 2π.

In most cases (and for the purposes of this chapter), sampling relates to band-limited images, i.e., images with maximum cyclic (or spatial) frequencies (Figure 1.5). The Nyquist (or Shannon) theorem, applicable to 1D signal processing, also applies, such that reconstruction of the original image is unique if the Nyquist (Shannon) sampling spatial rates are satisfied, that is,

$$\Delta x.2\omega_{x,\max} \leq 2\pi \qquad (1.39)$$

$$\Delta y.2\omega_{y,\max} \leq 2\pi \qquad (1.40)$$

This yields the maximum sampling periods of

$$\Delta x_{\max} = \frac{\pi}{\omega_{x,\max}} \qquad (1.41)$$

$$\Delta y_{\max} = \frac{\pi}{\omega_{y,\max}} \qquad (1.42)$$

Sampling with periods less than the maximum sampling periods results in image aliasing or wraparound artifacts.

1.17 Properties of Discrete Fourier Transforms

In a similar fashion to continuous FT, the discrete FT has corresponding properties, summarized below (Table 1.2).

Table 1.2. Listing of Some of the Most Important Discrete Fourier Transformations

Property	Transformation
Linearity	$FT[\alpha.i_1(m, n) + \beta.i_2(m, n)] = \alpha FT[I_1(m, n] + \beta FT[I_2(m, n)] = \alpha I_1(\omega_x, \omega_y) + \beta I_2(\omega_x, \omega_y)$
Shifting	$FT[i(m - m_o, n - n_o)] = I(\omega_x, \omega_y).e^{-j(\omega_x m_0 + \omega_y n_0)}$
Convolution	$FT[i_1(m, n) ** i_2(m, n)] = I_1(\omega_x, \omega_y).I_2(\omega_x, \omega_y)$
Separability	$FT[i_1(m, n).i_2(m, n)] = I_1(\omega_x).I_2(\omega_y).I_2(\omega_x), I_1(\omega_y)$
Parseval's theorem	$\sum\limits_{m=-\infty}^{+\infty} \sum\limits_{n=-\infty}^{+\infty} \left|i(m,n)\right|^2 dxdy = \dfrac{1}{4\pi^2} \int_{-\omega}^{+\infty} \int_{-\omega}^{+\infty} \left|I(\omega_x, \omega_y)\right|^2 d\omega_x d\omega_y$

Example 1.4

1. How will a typical MR image $I(x, y)$ change if the image is filtered by thresholding it at various levels κ?
2. A typical image is represented by matrix I. Compute the resulting image after application of a mean filter Φ.

ANSWER

1. Starting from a two-dimensional image, apply thresholding and mean filtering to compute resulting images. Assume that the choice of a threshold is defined by one of the following:

$$I_\kappa(x,y) = \begin{cases} 1 & if \ I(x,y) \geq \kappa \\ 0 & < \kappa \end{cases} \quad \text{(Ex 1.17)}$$

$$I_\kappa(x,y) = \begin{cases} 1 & if \ I(x,y) > \kappa \\ 0 & \leq \kappa \end{cases} \quad \text{(Ex 1.18)}$$

$$I_\kappa(x,y) = \begin{cases} 0 & if \ I(x,y) \geq \kappa \\ 1 & < \kappa \end{cases} \quad \text{(Ex 1.19)}$$

$$I_\kappa(x,y) = \begin{cases} 0 & if \ I(x,y) > \kappa \\ 1 & \leq \kappa \end{cases} \quad \text{(Ex 1.20)}$$

In accordance to the first definition:

$$I_k(x, y) = \begin{pmatrix} 0 & 1 & 1 & 0 & 0 & 1 & 0 \\ 0 & 1 & 1 & 1 & 0 & 0 & 0 \\ 1 & 1 & 1 & 1 & 1 & 1 & 0 \\ 1 & 1 & 1 & 1 & 1 & 0 & 1 \\ 1 & 1 & 1 & 1 & 1 & 0 & 0 \\ 1 & 1 & 1 & 1 & 1 & 1 & 1 \\ 1 & 1 & 1 & 1 & 1 & 1 & 1 \\ 1 & 1 & 1 & 1 & 1 & 0 & 1 \end{pmatrix} \quad \text{(Ex 1.21)}$$

If the second definition is obeyed:

$$I_k(x,y) = \begin{pmatrix} 0 & 1 & 1 & 0 & 0 & 1 & 0 \\ 0 & 1 & 1 & 1 & 0 & 0 & 0 \\ 1 & 1 & 1 & 1 & 1 & 1 & 0 \\ 1 & 1 & 1 & 1 & 1 & 0 & 1 \\ 1 & 1 & 1 & 1 & 1 & 0 & 0 \\ 1 & 1 & 1 & 1 & 1 & 1 & 1 \\ 1 & 1 & 1 & 1 & 1 & 1 & 1 \\ 1 & 1 & 1 & 1 & 1 & 0 & 1 \end{pmatrix} \qquad \text{(Ex 1.22)}$$

Similarly, if $\kappa = 1$, then:

$$I_k(x,y) = \begin{pmatrix} 0 & 0 & 0 & 0 & 0 & 0 & 0 \\ 0 & 1 & 1 & 0 & 0 & 0 & 0 \\ 0 & 1 & 1 & 0 & 1 & 1 & 0 \\ 1 & 1 & 1 & 1 & 1 & 0 & 0 \\ 1 & 1 & 1 & 0 & 1 & 0 & 0 \\ 1 & 1 & 1 & 1 & 1 & 1 & 0 \\ 1 & 1 & 1 & 1 & 1 & 0 & 1 \\ 1 & 1 & 1 & 0 & 1 & 0 & 0 \end{pmatrix} \qquad \text{(Ex 1.23)}$$

The result is the same for $\kappa = 2, 3, \ldots, 8$.

2. Depending on the choice of the filter type (2×2 or 3×3), the solutions are as follows.
   For a 2×2 median filter:

$$\Phi(x,y) = \begin{pmatrix} 1/4 & 1/4 \\ 1/4 & 1/4 \end{pmatrix} \qquad \text{(Ex 1.24)}$$

If the filter size is 3×3, then:

$$\Phi(x,y) = \begin{pmatrix} 1/9 & 1/9 & 1/9 \\ 1/9 & 1/9 & 1/9 \\ 1/9 & 1/9 & 1/9 \end{pmatrix} \qquad \text{(Ex 1.25)}$$

In the case of a 2×2 median filter:

$$I(x,y) = \begin{pmatrix} 5/4 & 11/4 & 7/4 & 1/4 & 1.4 & 11/4 & 0 \\ 3 & 22/4 & 13/4 & 1 & 11/4 & 9/4 & 0 \\ 4 & 6 & 15/4 & 2 & 13/4 & 10/4 & 1/4 \\ 17/4 & 26/4 & 17/4 & 2 & 1 & 1/4 & 1/4 \\ 17/4 & 26/4 & 21/4 & 4 & 11/4 & 1 & 1/4 \\ 17/4 & 26/4 & 25/4 & 25/4 & 4 & 7/4 & 3/4 \\ 14/4 & 19/4 & 4 & 4 & 11/4 & 1 & 3/4 \\ 5/4 & 1 & 3/4 & 1 & 3/4 & 1/4 & 1/4 \end{pmatrix} \qquad \text{(Ex 1.26)}$$

Solve in a similar fashion for the case of a 3×3 median filter.

Selected Readings

1. Macovski A. *Medical Imaging Systems*. Prentice Hall, 1983, Englewood Cliffs, NJ.
2. Jain AK. *Fundamentals of Digital Image Processing*. Prentice Hall, 1989, Englewood Cliffs, NJ, pp. 11–15.

② Fundamentals of Magnetic Resonance I: Basic Physics

2.1 Introduction

The physics of magnetic resonance imaging (MRI) were first described in 1946, and nuclear magnetic resonance (NMR) images have been used since 1973. The NMR phenomenon was originally described independently at Stanford University by Felix Bloch (Bloch 1946) and at Harvard by Edward Purcell, in 1946. Both received the Nobel Prize in Physics in 1952 for their work.

Magnetic resonance as a diagnostic modality and imaging technique exhibits various advantages compared to other commonly used alternative diagnostic radiological techniques, including the fact that it

- Uses no ionizing radiation

- Is known to have no biological hazards

- Produces images with high resolution

- Is a direct pulse acquisition technique

- Exhibits excellent soft tissue contrast

There are two approaches that can explain the physics of MRI:

- The quantum mechanical approach

- The classical NMR approach at a macroscopic level

Both treatments reduce to the same results, however, the mathematical complexity of the quantum mechanical approach is cumbersome. A succinct and simplified quantum mechanical approach follows, leading to the macroscopic description for explaining the phenomenon of NMR. The interested reader is referred to the excellent textbooks of Slichter (1990) and Abraham (1972) for an analytical description of the quantum mechanical explanation of NMR, and to the more recent textbooks by Callaghan (Callaghan 1993), Hornak (Hornak 2004), and others (Morris 1986, Freeman 1987, Hashemi 1998), for a description based on the classical approach.

2.2 Quantum Mechanical Description of NMR: Energy Level Diagrams

Important to the understanding of the underlying principles of MRI are the physical phenomena that link a moving charged particle and a magnetic field, that a moving charge gives rise to a magnetic field and that it experiences a force when moving within another magnetic field. Using the same analogy, if we consider the nucleus of an atom as a spinning cloud charge about its axis of rotation, a magnetic field or moment μ is produced, characteristic of such a nuclear magnetism, pointing along the spin axis. The magnetic response from an ensemble of atoms gives rise to a net magnetization vector M. The value of the magnetic moment takes only discrete values determined by the allowed values of angular momentum acquired, according to

$$\mu = \gamma \frac{h}{2\pi} \sqrt{I(I+1)} \qquad (2.1)$$

where γ is the gyromagnetic ratio (a constant characteristic of a spinning nucleus), h is the Planck's constant, and I is the spin quantum number. Placement of nuclear spins in an external magnetic field B_o allows them to take up discrete energy levels, with their magnetic moment becoming aligned with the direction of the external magnetic field (parallel or antiparallel at the lowest possible energy configuration (Stern and Gerlach 1922). In fact, the spin gyroscopic property actually forces them to precess at a small angle α with respect to the direction of the external field B_o (Figure 2.1).

For a nucleus with a spin quantum number I, there are $(2I + 1)$ different possible orientations of μ in the field. The solution of the Schrödinger equation using the wave Hamiltonian indicates that for a nucleus with $I = 1/2$, μ can take two values, with energies being determined by their magnetic quantum number m_i (that takes values $I, I - 1, ..., -I - 1, -I$). The energy difference of the states that the spins take is given by

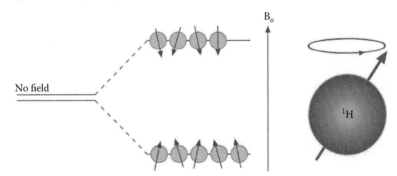

Figure 2.1. Left–middle: Nuclei occupy discrete energy levels in the presence of an external magnetic field. Right: Precession of ^1H nuclei at an angle α with respect to the direction of an external magnetic field, due to their gyroscopic properties.

$$\Delta E = \gamma \frac{h}{2\pi} B_o \Delta m_i \qquad (2.2)$$

Some nuclei, such as ^1H, ^{13}C, and ^{31}P, have a spin quantum number $I = 1/2$, and therefore the spins can exist in one of two possible states with magnetic quantum numbers $1/2$ and $-1/2$, and an energy difference between allowable energy states of $\Delta E = \gamma \dfrac{h}{2\pi} B_o$.

2.3   Boltzmann Statistics

At thermal equilibrium, most nuclear spins exist in the lowest energy state; however, at temperatures above absolute zero, there is an increasing number of nuclei at the higher energy state. The difference in the spin populations is given by the Boltzmann factor:

$$\frac{n_m}{n_{-m}} = e^{-\left(\frac{\Delta E}{kT}\right)} \qquad (2.3)$$

where n_{-m}, n_{+m} represent the number of nuclear spins at the low and high energy states, ΔE is the energy difference in the two states (Zeeman effect), k is the Boltzmann factor, and T is the absolute temperature. Therefore, a larger initial net magnetization vector M is obtained at higher field strengths and at lower temperatures.

These transitions (low to high energy levels and vice versa) are also subject to thermal perturbations and exchange between rotating spins within the sample. The latter occur due to rotational, translational, and vibrational motions that lead to inelastic collisions. In effect, the opposing processes that take place are

- The existence of the external static B_o aligns spins along the low energy state.

- Thermal fluctuations cause spins to move to a higher energy level.

- Thermal fluctuations distribute the protons randomly between the two states.

The outcome of such effects is given by the Boltzmann equation. At any particular instant, the excess number of nuclei in the low energy state is only about 1 part per million (ppm). This implies a stable thermal equilibrium between the two states and a net magnetization vector along the direction of the external field.

Transitions of spins from the lower energy level to a higher energy level require energy transfer. In NMR, such energy is provided to the spins via a weak external RF field at the resonant frequency, f_o, known as the Larmor frequency:

$$f_o = -\frac{\gamma}{2\pi} B_o \qquad (2.4)$$

2.4   Pulsed and Continuous Wave NMR

The resonance phenomenon can be observed experimentally by measuring the total absorbed energy of the system by sweeping the excitation frequency throughout the entire range of the electromagnetic spectrum. This is known as the continuous wave (CW) NMR.

Alternatively, pulsed NMR methods use low-power radio frequency pulses (with a fairly broad frequency bandwidth) to excite the spin system, and therefore modify the spin populations at the two energy states, thereby manipulating the net magnetization vector in a direction away from the direction of the externally applied magnetic field. During the relaxation phase, spins radiate the electromagnetic energy, which is detected by a radio frequency antenna, suitably placed on the transverse plane.

2.5 Spin Quantum Numbers and Charge Densities

Considering an atom $_X^A N$ with p protons and n neutrons, a nuclear mass number $A = p + n$ and an atomic $X = p$, an NMR signal can only exist for nonzero values of the spin quantum number I according to what is shown in Figure 2.2.

Nuclei (isotopes) such as ^{12}C, ^{10}O, and ^{40}Ca lead to no net magnetic field and are therefore invisible to NMR. Important nuclei metabolically are ^{1}H, ^{13}C, ^{31}P, ^{23}Na, ^{31}K, etc. (Please see Table 2.1.) In most of the remaining chapters of this book reference is made to proton MRI, that is, MRI of the hydrogen nuclei that exist mostly on water and lipids.

2.6 Angular Momentum and Precession

In the presence of an external magnetic field B, the net magnetic moment (or magnetization vector) M will experience a torque T governed by basic physics and given by

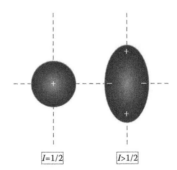

Figure 2.2. Nuclear charge distribution for nuclei with spin quantum number of $I = 1/2$ and $I > 1/2$.

Table 2.1. Spin Quantum Characteristics for All Atoms in the Periodic Table

Number of Neutrons (n)	Number of Protons (p)	Spin Quantum Number (I)
Even	Even	$I = 0$—no NMR phenomenon
Odd	Odd	I = Integral value—electric quadrupole—broad NMR lines
Odd	Even	I = 1/2, 3/2, 5/2, 7/2—high-resolution NMR (I = 1/2)
Even	Odd	$I > 1/2$—electric quadrupole—broad NMR lines

Note: The spin quantum number I is a measure of the charge distribution of the nucleus. For $I = 1/2$ the charge is evenly distributed over a sphere, whereas for $I > 1/2$ the charge distribution becomes asymmetric or ellipsoidal.

$$T = \frac{dJ}{dt} = M \times B \tag{2.5}$$

J is the angular momentum defined by

$$M = \gamma J \tag{2.6}$$

Using the analogy to spin quantum mechanics, J is the spatial average of the spin angular momentum operator. Therefore, the equation of motion of M about B_o over a narrow time period, in the presence of an external magnetic field, is given by

$$\frac{dM}{dt} = \gamma M \times B \tag{2.7}$$

This precession occurs at the Larmor frequency.

2.7 Overview of MR Instrumentation

See Figure 2.3.

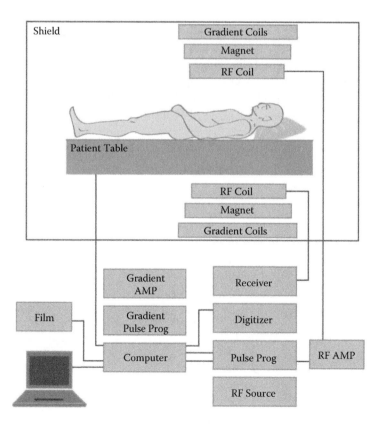

Figure 2.3. A schematic representation of all MR hardware and interconnections. Also shown are the directions of the externally applied magnetic field and the placement of the RF coil for signal detection.

2.8   The Classical View of NMR: A Macroscopic Approach

Less obvious but easier to comprehend is the classical description of MRI. Such an approach employs the net magnetization vector M from an ensemble of spins of a volume of tissue to describe the underlying physical phenomena.

2.8.1   The Net Magnetization Vector

Based on prior explanations, protons in the lower energy state are in excess from the number of protons in the higher energy state. The population difference in the two states increases at higher field strengths and at lower temperatures. This excess in spin populations in the lower energy level is the reason for the existence of a net magnetization vector M. Therefore, to understand and be able to design magnetic resonance imaging techniques, the equations of motion of $M(t)$ need to be considered and solved.

If B_o is the external static magnetic field along the z direction, then reference to Figure 2.4 leads to the initial components of the magnetization vector in the Cartesian coordinate system as

$$M_z = M_o \cdot \cos\alpha \tag{2.8}$$

$$M_x = M_o \cdot \sin\alpha \cdot \cos(\omega_o t + \phi) \tag{2.9}$$

$$M_y = M_o \cdot \sin\alpha \cdot \cos(\omega_o t + \phi) \tag{2.10}$$

where M_o is the net equilibrium magnetization vector, M_z is the longitudinal component, M_{xy} is the transverse component, φ is an arbitrary initial angle position of the magnetization vector component in the transverse plane, and α is the tip or flip angle. The resonant frequency or Larmor frequency ω_0 is defined by

$$\omega_o = 2\pi f_o = -\frac{\gamma}{2\pi} \cdot B_o \tag{2.11}$$

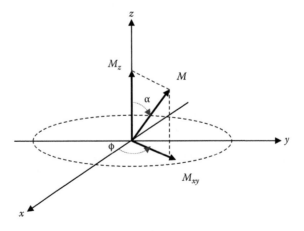

Figure 2.4. The net magnetization vector M as it precesses about the external static magnetic field B_o. (Adapted from a figure by Professor E. McVeigh. With permission.)

For protons at a field strength of 1.5 Tesla, $\gamma/2\pi = 42.58$ MHz/Tesla. Notably, choice of a tip angle of 90° elicits the maximum value of the transverse component of the magnetization vector upon first excitation.

Endorsing Euler's convention to represent the transverse component of the magnetization vector as a rotating phasor, M_{xy} can be written as

$$M_{xy} = M_o \sin\alpha . e^{i(\omega_o t + \phi)} \qquad (2.12)$$

2.9   Rotating Frame and Laboratory Frame

Direct observation of the MRI signal (as described in the previous section) is often done at the laboratory frame of reference, from the x, y, and z axes of the Cartesian coordinate system. Resonance, even at relatively low field strengths (e.g., 1.5 T) occurs at frequencies at a few MHz. Frequency differences in the spin population that may occur as a result of local changes are small. The rotating frame of reference is therefore introduced at axes x', y', and z'. This convention provides a convenient way of observing the spin system, as in a frame that is rotating with the same frequency as the resonant frequency. Practically, and following techniques often used in communications, a phase-sensitive coherence detection receiver is used to detect and demodulate the signal at baseband and subsequently represent it relative to 0° and 90° (out of phase), that is, at quadrature. Quadrature detection will be revisited when spectroscopy and spectroscopic imaging are discussed.

2.10   RF Excitation and Detection

In the presence of a strong external magnetic field B_o, the net magnetization simply precesses about the static field at the natural frequency (Larmor frequency). Application of a weak, circularly polarized B_1 RF oscillating field (with frequency at right angles with respect to the equilibrium magnetization vector M_o) will force M_o to undergo nutation (Figure 2.5).

The amount of nutation, or tip angle, α, away from the z axis can be changed by changing the duration or amplitude of the external magnetic field B_1 (applied with a separate RF coil). A 90° pulse is known as an excitation phase, and places the magnetization on the transverse plane. Equivalently, a 180° pulse places the magnetization on the negative z axis. In general,

$$\alpha = -\int_0^t \gamma . B_1(t)\, dt \qquad (2.13)$$

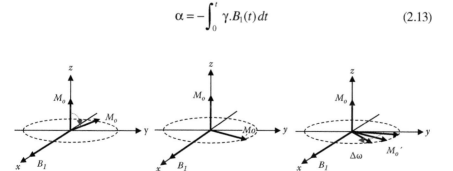

Figure 2.5. Nutation of the magnetization vector M upon irradiation with a circularly polarized RF pulse perpendicular to the z axis (diagrams depict vectors in the rotating frame).

For a constant width × amplitude B_1 pulse, the tip angle becomes

$$\alpha = -\gamma.B_1.\Delta t_{rf} \tag{2.14}$$

where Δt_{rf} represents the time duration of the RF pulse in milliseconds. From Faraday's law of induction, any changing magnetic field produces an electromotive force (EMF), which is detected as a voltage in the receiving coils. The amplitude of the induced RF voltage is minute (only of the order of a few μV), and it is amplified before being sent to the receiver for demodulation and sampling.

2.11   Molecular Spin Relaxation: Free Induction Decay

2.11.1   Relaxation Mechanisms

After the application of the excitation pulse, the transverse magnetization precesses in the transverse plane and decays exponentially with time as a result of the intrinsic relaxation process in tissue, characterized by a constant known as T_2. This parameter is also known as the transverse relaxation constant (or spin–spin relaxation constant), and the signal evolution with time is known as the *free induction decay* (FID) (Figure 2.6).

Mathematically, the transverse magnetization vector can be expressed as

$$M_{xy} = M_o \sin\alpha.e^{i(\omega_o t + \phi)}.e^{-t/T_2} \tag{2.15}$$

The longitudinal magnetization relaxes to equilibrium according to a totally independent physical process than for the transverse relaxation, characterized by a constant known as T_1. The T_1 relaxation process characterizes the transitions of excited spins from the higher energy level state (spin-down state). It is commonly known as the spin–lattice relaxation time. Effectively, T_1 describes the rate at which the spin system returns back to thermal equilibrium. Mathematically, the evolution of the longitudinal magnetization vector will obey the solution of the ordinary differential equation:

$$\frac{dM_z(t)}{dt} = \frac{[M_o - M_z(t)]}{T_1} \tag{2.16}$$

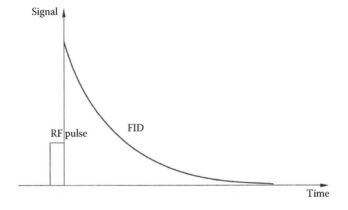

Figure 2.6. Representation of a free induction decay following a 90° RF pulse.

Given an α pulse and the initial condition under which the magnetization vector is fully relaxed, the solution becomes

$$M_z(t) = M_o(1 - e^{-t/T_1}) + M_o.\cos\alpha.e^{-t/T_1} \qquad (2.17)$$

Example 2.1

Transform the Bloch equation into the rotating frame.

ANSWER

The Bloch equation can be written as

$$\frac{dM}{dt} = \gamma M \times B - R[M - M_o] \qquad (Ex\ 2.1)$$

- $M \times B$ and $R[M' - M_o]$ apply in any frame.
- Assume that the rotating frame rotates with angular frequency Ω with respect to the lab frame, and that M' is the magnetization vector in the rotating frame (B' is B in the rotating frame).

From classical mechanics:

$$\frac{dM}{dt} = \frac{dM'}{dt} + \Omega \times \frac{dM'}{dt} \qquad (Ex\ 2.2)$$

Using the above two equations:

$$\frac{dM'}{dt} = \gamma M' \times B' - R[M' - M_o] - \Omega \times \frac{dM'}{dt} \qquad (Ex\ 2.3)$$

2.11.2   Mechanism of Relaxation Processes: T_1 Relaxation

Relaxation processes and the molecular environment will be dealt with in detail in Section 2.11.3. As the water molecule in the molecular milieu undergoes translational, vibrational, and rotational motion in the presence of the external field, each of the two water protons remains, nearly all of the time, independently aligned parallel or antiparallel to B_o.

So, the dipole magnetic field that a proton produces in the vicinity of its partner proton ranges from nearly zero to a few mT. As the molecule tumbles, there is a magnetic field generated, having a maximum amplitude of approximately 0.4 mT. This phenomenon is intramolecular and not intermolecular, and the dipole strength falls as $1/r^6$, where r represents the spatial distance away from the proton nucleus. The effect on neighboring molecules thus becomes weaker with distance.

If the water molecules get exposed to B_o and B_1 at the Larmor frequency, the dipole magnetic field produced by proton a and felt by its partner proton (b) will also fluctuate at the Larmor frequency with a maximum amplitude of 0.4 mT. Such a field can cause proton spin-state transitions. So even if there is no energy supplied from external fields, a proton can still be raised to the higher energy level or equivalently "thrown out" to a lower energy state.

Therefore, the longitudinal relaxation time T_1 is largely determined by the oscillating fluctuation in the local magnetic fields (at Larmor frequency). It appears that T_1 also depends on the degree to which water molecules within a cell are bound to intracellular macromolecules. The amount of water and the binding of water molecules are in turn dependent on the cell type and its physiological status. Tissue T_1 is therefore affected by the tissue composition and is modulated significantly in different pathophysiological states. It is for this reason that a major objective in MRI studies included the detection of variations in the signal intensity as a consequence of T_1 variations that inherently associate with the molecular environment and its changes.

2.11.3   Mechanism of Relaxation Processes: T_2 Relaxation

If the spin–lattice relaxation mechanism was the only relaxation mechanism, the transverse magnetization in the xy plane would fall off with the same characteristic time T_1. This is only true for nonviscous fluids (with T_1 being approximately equal to T_2). In simpler terms, in such an environment there are not many spin–spin interactions, and therefore the secular contribution to T_2 is almost nonexistent (see below). However, with solids, viscous fluids, and tissues, the FID falls off much faster than if it were driven by T_1 alone. The time constant T_2^* (known also as T_2-star) characterizes this process. If the water molecule is tumbling freely, as in pure water, such fields will be changing too fast to have an appreciable effect on the Larmor frequency of precession of protons.

If, however, the water molecular motion is significantly slowed down, then each proton will be subjected to a relatively slowly varying static dipole field from its partner proton—sometimes adding and sometimes subtracting from its own dipole, thereby causing the proton to precess faster or slower than the Larmor frequency. This effect will lead to loss of coherence, and hence to a decaying transverse magnetization.

External field inhomogeneities due to the external magnet are often present. The FID signal is obtained due to the net transverse magnetization that arises as a result of the precessing of the spins in coherence. If, however, the external field is not uniform (because of the plane of the imperfect magnet design), then the individual protons experience local magnetic fields that are slightly different, and they precess at slightly different frequencies. At any instant in time, approximately half the protons will be precessing at faster angular velocities than the average proton, and the rest with slower angular velocities.

Therefore, after an initial 90°, the spins will lose their initial phase coherence, and the spin interactions will fan out and disperse. Eventually the magnetic field moments will cancel out, thus leading to a zero net magnetic field. Longitudinal relaxation (involving transitions between the lower and higher spin states) is only one of the two processes responsible for the spin dephasing characterized by T_2, the other being the secular contribution.

2.12 T_1 and T_2 Measurements

2.12.1   Measurement of T_1 and T_2

There are multiple ways to measure T_1 or T_2 values of biological tissues, but the most commonly used methods are the saturation recovery (or inversion recovery) for measuring T_1 and the spin–echo for measuring T_2.

2.12.2   Saturation Recovery: T_1 Measurement

A 90° pulse is known as a saturation pulse and drives the spin system away from equilibrium. Explicitly, at the onset of excitation ($t = 0$), the 90° pulse tips the magnetization vector in the transverse plane, inducing a voltage in the detection coils. The longitudinal magnetization is then left to recover for $t =$ TI ms (known also as the inversion time), and then another 90° pulse interrogates the magnetization vector. The system is then allowed to return to thermal equilibrium (Figure 2.7).

The entire process is repeated several times with different choices of TI. In a T_1-weighted experiment, maximization of the contrast between tissues based on their T_1 recovery is achieved (e.g., lipid yields much higher signal than white and gray matter in the brain).

2.12.3   Spin–Echo: T_2 Measurement

The presence of local field inhomogeneities affects the transverse relaxation process. In effect, the signal decay becomes faster, characterized by a parameter $T_2{}^*$, defined by

$$\frac{1}{T_2^*} = \frac{1}{T_2} \pm \Delta\, B_o \qquad (2.18)$$

where ΔB_o represents the local variations in the external static field due to local inhomogeneities. Spin–echo is a technique introduced by Hahn (1950) and is immune to alterations in $T_2{}^*$ effects that arise from the inhomogeneity of the static field (Figure 2.8).

2.12.4 T_1 and T_2 in Solids and Liquids

Solids have relatively fixed molecules with external magnetic fields causing significant local variations in the value of the magnetic field around the lattice protons. In a liquid (or soft tissue), the local magnetic fields from neighboring molecules fluctuate rapidly as the molecules move about, and the end result is a small net contribution to the net magnetic field.

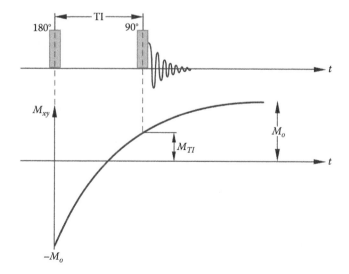

Figure 2.7. An inversion recovery experiment for measuring T_1.

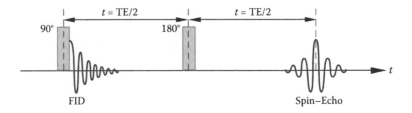

Figure 2.8. A spin–echo experiment for measuring T_2.

T_1 is much longer for solids than for liquids. In solids, there is not a significant transfer of energy between the spinning molecules and the lattice. T_2 is much shorter for solids than for liquids. Due to significant local variations in the value of the magnetic field around protons, spinning molecules interact significantly, hence losing their coherence. In a liquid, despite the fact that molecules are free to move rapidly, their net local magnetization averages very quickly to zero and contributes little to the spin–spin interactions. Thus, the dephasing between the spins is less, and T_2 is larger.

2.13   Relaxation Times in Biological Tissues

2.13.1   Liquid State: Small and Large Macromolecules

If the fluid contains small molecules, the rate of movement or molecular tumbling occurs at much higher frequencies than the Larmor frequencies used in NMR, and thus there is an inefficient energy exchange. Thus, T_1 is relatively large in value in liquids, with small molecules within them (such as in the cerebrospinal fluid (CSF)). Conversely, most human body tissues are composed of macromolecules with much larger molecular sizes, such as proteins, fats, and complex sugars, which make their T_1 values much shorter. For example, white matter has shorter T_1 values than gray matter since its protons are bound to the much larger myelin (fatty) molecules.

2.13.2   Clinical Correlations

In magnetic resonance, the emitted signal from a particular tissue is a result of the tissue T_1 and T_2 values. In constructing a clinical image, a radiologist is trained to view and diagnose images that are either T_1, T_2, or proton density weighted. In a T_1-weighted image, tissues with low T_1 values will be displayed as bright elements and tissues with low T_2 as dark areas. Thus, a T_1- and a T_2-weighted image for the same exact anatomical area can look quite dissimilar.

An MRI image is not a direct voxel-to-voxel map of the relaxation times T_1, T_2, or the proton density. It is rather a map of the signal distribution in each voxel due to the magnetization vector. The magnetization does depend on T_1, T_2, and proton density, but also on the type of imaging acquisition scheme (the way the magnetization vector is manipulated with time, i.e., the equation of motion characterizing $M(t)$, and other imaging parameters). T_1 and T_2 are of prime importance for MRI, since they affect, among others, tissue contrast.

Selected Readings

1. Abraham A. *The Principles of Nuclear Magnetism.* Oxford University Press 1972, Oxford.
2. Callaghan PT. *The Principles of Nuclear Magnetic Resonance.* Oxford University Press, 1993, Oxford.

3. Ernst R, Bodenhausen G, Wokaun A. *Principles of Nuclear Magnetic Resonance in One and Two Dimensions.* Oxford University Press, 1987, Oxford.

4. Hahn EL. Spin–Echoes. *Physical Review* 1950; 80(4):580–594.

5. Sprawls P, Bronskill MJ. *The Physics of MRI: 1992 AAPM Summer School Proceedings.* American Association of Physicists in Medicine, 1993, Woodbury, NY.

6. Stark D, Bradley W. *Magnetic Resonance Imaging.* Mosby, 1999, St. Louis, MO.

7. Morris PG. *Nuclear Magnetic Resonance Imaging in Medicine and Biology.* Oxford University Press, 1986, Oxford.

8. Hornak JP. *The Basics of MRI.* 2004. http://www.cis.rit.edu/htbooks/mri/.

9. McRobbie DW, Moore EA, Graves MJ, Prince MR. *MRI—From Picture to Proton.* Cambridge University Press, 2003, Cambridge, pp. 164–188.

③ The Molecular Environment and Relaxation

3.1 Introduction

In the introductory chapters it has been shown that for $I = 1/2$ nuclei, there are two discrete energy levels to be taken up by spins, with the majority of the spins populating the lower energy level (in excess to the upper energy level population by approximately 1 ppm). After a saturation recovery or an inversion recovery experiment, the Boltzmann distribution is perturbed. Return of the spin population back to equilibrium occurs via transition of nuclei from the higher energy level back to the low energy level via either a spontaneous (optical spectroscopy) or a stimulated emission. The probability for a spontaneous emission in the radio frequency (RF) range of frequencies is nonsignificant. Such transitions, accompanied by a loss of quanta of energy ΔE, restore the Boltzmann distribution, and the process is known as spin–lattice relaxation.

In this chapter spin–spin and spin–lattice relaxations are revisited to explain the phenomenon of relaxation at the molecular level.

3.2 Biophysical Aspects of Relaxation Times

There have been numerous publications and work in the investigation of relaxation times of water in tissue that span several decades. Fundamental to the field of nuclear magnetic resonance (NMR) was the publication of Bloembergen, Purcell, and Pound (often referred to as the BPP publication) that was published in 1948 (Bloembergen et al. 1948) that discussed the relation of molecular motions to relaxation in aqueous solutions. The theory behind such work was by itself inadequate in explaining relaxation in tissues. Critical to the understanding of relaxation and relaxation changes in tissue is cellular and molecular compartmentalization. The cellular milieu can be thought of as an ensemble of different compartments comprising molecules and other species that exchange between these at different rates. The simplest case is that characterized by fast exchange, that is, exchange of molecular moieties in distinct molecular compartments at timescales that are much smaller than the relaxation values of these moieties. Another example is that of restricted diffusion that matches a number of biologically relevant cases, including water diffusion along fiber structures, and diffusion of ions downstream or upstream of concentration gradients. This has been an active area of research, and more work is envisaged in the future. Pharmacological studies and the latest developments in molecular imaging benefit directly from such considerations.

3.3   Spectral Density and Correlation Times

Perturbations of the spin population by irradiation of a sample with RF power leads to net transverse magnetization. Interaction between protons leads to spin exchange that causes random spin dephasing of the transverse magnetization, characterized by the spin–spin relaxation constant, known as T_2. This constant is characteristic for the tissue. In a similar fashion, the set of mechanisms that give rise to fluctuating magnetic fields (including vibrational, rotational, translational, and other molecular motions) cause spins to dissipate energy by returning to a lower energy state, a transition characteristic by the longitudinal relaxation time, known as T_1. In justification of these important mechanisms for the phenomenon of NMR, a detailed discussion on the nature of the molecular motions and reasons for T_1 and T_2 relaxation follows (Bottomley 1984).

Every nucleus in tissue is surrounded by many other molecules and particles with magnetic moments that interact with each other. Dipole-dipole interactions are important at the spatial scales and separations of neighboring dipoles present in biological tissue. As a result of the random rotational, vibrational, and translational motions of such molecules, the local magnetic field surrounding a nucleus fluctuates constantly in both space and time.

In any material, each proton is exposed to a spectrum of locally changing magnetic fields, characterized by the so-called spectral distribution or spectral density function, $J(\omega)$, for a range of angular frequencies ω. Figure 3.1 shows typical plots for the spectral density function for a solid (Purcell 1946), a liquid, and semisolids, materials in an intermediate category, such as biological tissues.

Molecules, complexes, hydrated ions, or macromolecular assemblies rotate and tumble at all frequencies, as characterized by their rotation times or correlation time, τ_c, that is, the inverse of the period of time taken for molecular rotation, tumbling, and translation. Biological tissue closely resembles the relaxation properties of a semisolid.

3.4 T_1 and T_2 Relaxation

The other major mechanism that leads to spin–lattice relaxation includes (in addition to the magnetic dipole-dipole coupling between nuclear magnetic moments and the surrounding molecules) the coupling of protons to magnetic fields generated by atomic electrons (especially for atoms or molecules with unpaired electrons). Such effect is more prominent when agents such as paramagnetic contrast agents are present in the molecular environment. Such compounds are associated with unpaired electrons that have a magnetic moment 1000 times more than that of the nucleus. A paramagnetic agent thus dominates the spin-lattice relaxation of nuclei in its neighborhood (such as Mn^{2+} complexes).

In contrast to T_1, T_2 is an independent relaxation process. It is attributed to the local perturbations of the magnetic field due to the magnetic dipoles of

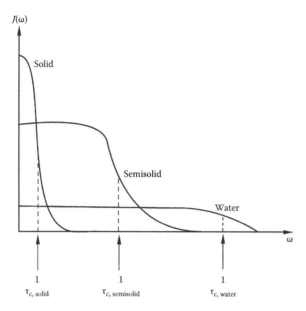

Figure 3.1. Spectral density function for solids, liquids, and semisolids. (Redrawn from Bronskill, M. J., and Graham, S., NMR Characteristics of Tissue, in *The Physics of MRI: 1992 AAPM Summer School Proceedings*, American Association of Physicists in Medicine, Woodbury, New York, 1993.)

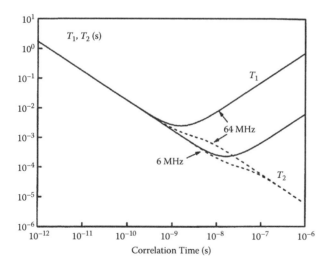

Figure 3.2. The BPP relationships of relaxation times T_1 and T_2 and correlation time τ_c. (Reproduced from Bronskill, M. J., and Graham, S., NMR Characteristics of Tissue, in *The Physics of MRI: 1992 AAPM Summer School Proceedings*, American Association of Physicists in Medicine, Woodbury, New York, 1993. With permission.)

the tumbling neighboring molecules and due to local magnetic fields generated from electrons that are additive or opposing to the external magnetic field B_o. Such a spatial and temporal variation of the static field as experienced locally leads to a dispersion in the precessional frequencies of the spins, causing them to "fall out" of coherence (Figure 3.2).

The equations for the transverse and longitudinal relaxation times can be derived as a function of the spectral density function (under certain assumptions), according to

$$\frac{1}{T_1} = \frac{9}{8}\gamma^4 \left(\frac{h}{2\pi}\right)^2 [J(\omega) + J(2\omega)] \tag{3.1}$$

$$\frac{1}{T_2} = \frac{3}{4}\gamma^4 \left(\frac{h}{2\pi}\right)^2 \left[\frac{3}{8}J(0) + \frac{15}{4}J(\omega) + \frac{3}{8}J(2\omega)\right] \tag{3.2}$$

$$J(\omega) = \frac{\tau_c}{[1 + (\omega_{io}\tau_c)^2]} \tag{3.3}$$

$$J(2\omega) = \frac{\tau_c}{[1 + (2\omega_{io}\tau_c)^2]} \tag{3.4}$$

where γ is the gyromagnetic ratio, and h is the Planck's constant. Notice that the expressions for the transverse and longitudinal relaxation times (or rates) are similar; the expression for T_2 has an additional term that characterizes the enhanced dephasing in the transverse plane due to the static ΔB_o perturbation in the vicinity of each proton. Notice also that only the fluctuations that occur at the resonant frequency, ω_o, and at twice the resonant frequency, $2\omega_o$, induce spin transitions that lead to longitudinal relaxation. The amplitudes of the spectral density function at such frequencies, $J(\omega_o)$, $J(2\omega_o)$, are thus a measure of the proportion of molecular interactions in tissue that induce longitudinal relaxation.

So, for liquids with a very short correlation time, $J(\omega_o)$ is low and T_1 is large (e.g., cerebrospinal fluid (CSF)). For tissues that have a correlation time comparable to the resonant frequency, the amplitude of the spectral density function is large and the relaxation process fast. In contrast, solids exhibit slow motion and hence have an ultra-fast relaxation. Additionally, such equations also explain why in liquids T_1 and T_2 are approximately equal ($J(\omega_o) \approx J(0)$) and $T_2 \ll T_1$.

3.5 Quadrupolar Moments

For spins with spin quantum numbers $I > 1/2$ the charge density around the nucleus is asymmetric, leading to quadrupolar moments that allow multiple energy transitions and therefore multiple relaxations. Irrespective of whether the nucleus is hydrated or not, the molecular complex will experience asymmetric electric gradients in its tumbling motion in the molecular milieu. A characteristic example includes nuclei with $I = 3/2$, such as the sodium or fluorine nuclei. Relaxation is characterized by a slow and a fast longitudinal and transverse relaxation process, with a pool contribution of 40–60%, analogous to the proportion of quantum transitions. The fast components of such relaxation components are so fast that they are sometimes undetected in commercially available scanners. Minimization of echo times by reducing amplifier blanking delays, slew rates, and for RF, pulse widths sometimes addresses the problem.

3.5.1 Biophysical Properties of Quadrupolar Nuclei

Quadrupolar nuclei have a spin quantum number of 3/2 and possess four energy levels. They also possess an asymmetric nuclear charge distribution that ultimately generates the quadrupolar moment. The moment interacts with the electric field gradients ΔE set up by the asymmetric distribution of dipoles surrounding the nuclei (Figure 3.3). In cases where spherical symmetry exists, no net electric field gradients exist at the site of the nucleus, and therefore the quadrupolar interaction is zero.

Transitions between different energy levels are allowed between two energy levels only if the difference in the spin quantum numbers (at the two different energy levels) is unity. The energy level diagram of the quadrupolar nucleus with $I = 3/2$ is depicted in Figure 3.4. If one considers *only* the Zeeman interaction, and in the absence of the quadrupolar interaction, the energy difference between adjacent levels is the same, and all transitions occur at the same frequency, resulting in a single NMR spectral line.

In order to explain the line shape of the quadrupolar nucleus resonance, both dipolar and quadrupolar interactions need to be considered.

3.5.1.1 Dipolar Interactions

Quadrupolar nuclei in aqueous solutions are surrounded by water molecules (for example, the sodium nucleus is associated with a hydration number of 6). Each of the water molecules has a dipole moment (Figure 3.5), and thus at the quadrupolar nucleus site there exists a fluctuating local magnetic field, experienced by the nucleus due to the motion of the hydrated complex within the solution.

In biological systems, there are also additional fields that need to be considered, due to protons or proteins or macromolecular complexes. The effective field produced by a proton at a distance r is given by

$$B_{local} = \frac{\mu_p}{r^3}(3\cos^2\theta - 1) \tag{3.5}$$

where μ_p is the proton magnetic moment, and θ is the angle between the external static magnetic field and the axis of symmetry (molecular axis). The line

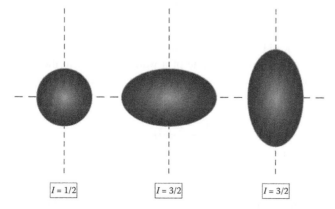

Figure 3.3. Examples of nuclear electric charge distributions for nuclei with different spin quantum numbers I.

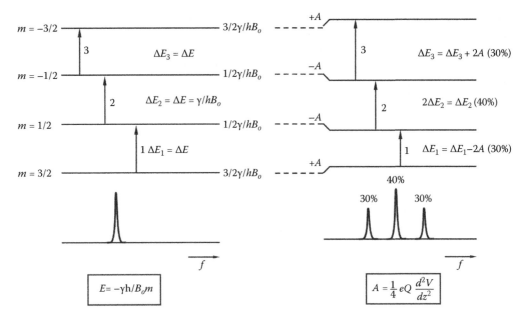

Figure 3.4. Quadrupolar splitting for a $I = 3/2$ nucleus in the (left) absence and (right) presence of an external magnetic field.

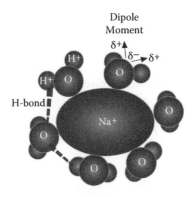

Figure 3.5. The hydrated sodium complex.

broadening produced by this dipolar interaction (T_2 relaxation) is primarily determined by either the limited lifetime of a water molecule in the hydration shell or the distribution of correlation times of rotation of the hydration complex (τ_c) (Bottomley et al. 1984), whichever is shorter, according to

$$\frac{1}{T_2} = \frac{2}{3} n \gamma^2 B_{local}^2 \tau_c \tag{3.6}$$

3.5.1.2   Quadrupolar Interactions

The quadrupolar interaction depends on the symmetry of the dipoles around the quadrupolar nucleus (referred to as the molecular complex) and the orientation of the symmetry axis (distribution of the polar water molecular around the quadrupolar nuclear complex) of the molecular complex relative to the external magnetic field. The energy of the four levels, given an axial symmetry, is given by

$$E(m) = E(m,0) + \frac{1}{16}e^2qQ(4m^2 - 5)(3\cos^2\theta - 1) \tag{3.7}$$

where $E(m, 0)$ is the Zeeman energy for level m, eq is the field gradient, eQ is the quadrupolar moment, and e is the electron charge.

As shown in Figure 3.4, all three energy transitions (in the presence of quadrupolar interactions) occur at three different frequencies. The position of the centerline is not affected by the quadrupolar coupling. The transitions 1, 3 lie on either side of the central NMR peak and are displayed by $\Delta\gamma = \pm 1/2e^2qQ$. Quantum mechanical calculations (Springer 1987) show that the centerline has 40% intensity compared to the sidelines, which have 30% intensity each. Since there is a dependency of the transition energies (due to quadrupolar interactions) on the orientation of the molecular symmetry axis, the frequency shift of the outer lines varies in a cosinusoidal manner with respect to the central line and is determined by

$$\Delta\gamma = \frac{1}{2}e^2qQ(3\cos^2\theta - 1) \tag{3.8}$$

Different splittings are therefore observed for different θ values based on the instant angle difference of the symmetry axis of the molecular complex with the magnetic field. Maximum splitting occurs when $\theta = 0°$, $180°$, and minimum splitting when $\theta = 55°$ (when the outer lines coalesce into the centerline). Thus, the positions of the outer lines are distributed over a frequency range from 0 to $\pm e^2qQ$ from the centerline. Since the area under these outer lines is constant, a frequency spread in their positions implies a reduction in the peak height. In addition, if there exists spatial heterogeneity in the molecular environment, the quadrupolar interaction eQ also varies, which leads to further broadening of the outer lines, beyond detection. This is a static effect, and the resulting broadening of the outer lines is referred to as heterogeneous broadening (Gupta et al. 1994) (Figure 3.5).

If there are molecular motions, they can cause fluctuations in the electric field gradients. Such fluctuations reduce the average value of the quadrupolar splitting, and in the limit of fast fluctuations (small correlation time), the quadrupolar splitting averages to zero, resulting in a single spectral line. In this case, the transitions occur at the same frequency as the central transition, because the average quadrupolar coupling is zero, thus leading to extremely short relaxation times, and very broad lines. Up to 60% of the signal is again lost, and this effect is known as homogeneous broadening.

3.5.1.3 Quadrupolar Effects on Relaxation Times

The T_1 and T_2 relaxation times are primarily influenced by the fluctuating electrical quadrupolar interactions. The assumption that the relaxation processes are primarily determined by rotational tumbling, with a single correlation time, is thus invoked. According to the theory proposed by Hubbard (1970), two components exist for T_1 and T_2 for quadrupolar nuclei:

$$\frac{1}{T_{1,I}} = \left(\frac{e^2qQ}{10}\right)\left(1 + \frac{\eta^2}{3}\right)\tau_c\left[\frac{1}{1 + \omega_0^2\tau_c^2}\right] \tag{3.9}$$

$$\frac{1}{T_{1,II}} = \left(\frac{e^2 qQ}{10}\right)\left(1+\frac{\eta^2}{3}\right)\tau_c\left[\frac{1}{1+4\omega_0^2\tau_c^2}\right] \tag{3.10}$$

where τ_c is the correlation time (which characterizes the duration of the position and orientation of a given molecule as it experiences an electric field gradient), ω_o is the resonant frequency, and η is defined by

$$\eta = \frac{\partial^2 V_{xx}/\partial x^2 - \partial^2 V_{yy}/\partial y^2}{\partial^2 V_{zz}/\partial z^2} \tag{3.11}$$

where V represents the electrical potential of the molecular complex. $T_{1,I}$ is not readily detectable in biological tissue since it accounts for only 20% of the total signal. The two transverse relaxation times are given by

$$\frac{1}{T_{2,I}} = \left(\frac{e^2 qQ}{20}\right)\left(1+\frac{\eta^2}{3}\right)\tau_c\left[\frac{1}{1+\omega_0^2\tau_c^2}+\frac{1}{1+4\omega_0^2\tau_c^2}\right] \tag{3.12}$$

$$\frac{1}{T_{2,II}} = \left(\frac{e^2 qQ}{20}\right)\left(1+\frac{\eta^2}{3}\right)\tau_c\left[1+\frac{1}{1+4\omega_0^2\tau_c^2}\right] \tag{3.13}$$

where $T_{2,II}$ represents the fast component (also denoted as $T_{2,f}$), which accounts for 60% of the total signal. $T_{2,I}$ is the slow component, commonly referred to as $T_{2,s}$. In conditions of fast motional tumbling, $\omega_0\tau_c \ll 1$, which leads to

$$\frac{1}{T_1} \approx \frac{1}{T_2} \tag{3.14}$$

a typical condition encountered in an aqueous solution ($\tau_c \sim 10^{-14}$ s). For the required frequency range of 10–100 MHz (typical clinical field strengths of 1.5 T), the condition $\omega_0^2\tau_c^2 \gg 1$ implied correlation times of the order of 10^{-14} to 10^{-18} s, much shorter than those in aqueous solution.

In biological tissue, the net correlation time for the quadrupolar nuclear complex will be the spatio-temporal density average of the rapidly tumbling complex, with the quadrupolar nucleus undergoing quadrupolar interactions. The relative shortening of the T_2 and corresponding broadening of the resonance line width for the fast fraction will depend on the strength of the quadrupolar interactions. It is possible that the fraction associated with fast tumbling may not be visible by MRI because of the difficulty in detecting short T_2 components in periods where spatially localized gradients are normally applied.

Care must be taken not to assign the fractions of slow and fast T_2 nuclei to the respective contributions from the intracellular and extracellular fractions, because the two components are the result of nuclear quadrupolar interactions with macromolecular charges that may occur both inside and outside the cell.

3.5.1.4   NMR Visibility of the Sodium Nucleus

Cope (1967), using continuous NMR, showed that sodium signals observed from muscle, kidney, and brain samples accounted for only 30–40% of the total sodium content in these tissues. Such observations were interpreted by postulating the existence of two populations of sodium ions: a bound population (to macromolecules), accounting for 60% of total sodium, and a free population, accounting for 40% of total ions in solution. The idea behind this postulate was that the bound ions have long correlation times, leading to short T_2 values, causing the NMR lines to broaden beyond detection. The observed NMR line was thought to originate from the free ions, and therefore has only 40% of the total intensity.

This hypothesis was initially supported by other independent studies and by the presence of two transverse relaxation times (Berendsen and Edzes 1973; Cope 1970; Shporer and Civan 1972; Chang and Woessner 1978). What was surprising was the fact that although the tissue studies were from various tissue parts (with different concentrations of macromolecules), they all had approximately the same concentration of free sodium. It was then that Shporer and Civan (1972) pointed out that these observations could be explained on the basis of a single population of sodium ions with quadrupolar coupling. Other evidence against the two-population compartmentalization theory was provided from the studies of Berendsen and Edzes (1973). They observed a single spin–lattice relaxation time even in the presence of two spin–spin relaxation times. If two populations were present, then two spin–lattice relaxation times would be expected.

By invoking the quadrupolar coupling, it was possible to explain why only 40% of the sodium was visible. Only the 40% associated with the central transition of the quadrupolar-coupled ^{23}Na may thus be observed. If the remaining 60%, which resides in the outer lines, is broadened, then it may not be detected. In biological tissues, the sodium complexes are randomly oriented and are in constant motion. Thus, the broadening of the outer transition may arise from both the homogeneous and the heterogeneous mechanisms described earlier. Although this quadrupolar interpretation has gained wide acceptance, the precise nature of the interaction is not clear. It could arise from either:

1. A rapid exchange between a small fraction of bound ions with the bulk of free sodium ions

2. A homogeneous population of sodium ions with some sort of ordering of macromolecules within the cell

However, the former hypothesis that rapid exchange between two homogeneous fractions of sodium ions might be responsible is unattainable because calculations indicate that in such a case the bound fraction would be fairly small, on the order of approximately 1% (Berendsen and Edzes 1973).

3.5.1.4.1 Model I The simplest model that would account for the 60% of intracellular sodium invisibility or broadening includes the presence of a single homogeneous population of intracellular Na$^+$, all of which would be immobilized. Previous work dismisses this hypothesis (Shporer and Civan 1974).

3.5.1.4.2 Model II The second simplest model involves the presence of a single population of ^{23}Na that undergoes rapid exchange between the immobilized and free species of Na$^+$ within an isotropic medium. Evidence shows that the amount of immobilization is very small, and thus only a few percent of the signal would be lost due to such an effect (Shporer and Civan 1974). On the other hand, the nuclear quadrupolar effect might arise from ordering, rather than from immobilization of Na$^+$.

3.5.1.4.3   Model III   A condensed phase of counterions exists on the surface of charged macromolecules. Such a layer could potentially modify the electric field gradient imposed on the sodium nucleus. Such macromolecules are usually incorporated into membrane structures or other intracellular organelles. In such an environment, the electric field gradient imposed on the Na^+ nuclei would not average out to zero with time, but would lead to an effective nuclear quadrupolar interaction. Calculations show (Shporer and Civan 1974) that if one assumes each Na^+ spends a residence time of τ_m within each domain, the lower limit for the extent of each domain is of the order of 100 Å.

Exchange of sodium between adjoining domains is driven by diffusion. If diffusion were to occur very slowly, this would explain the presence of the satellite peaks in the Na^+ spectrum. Note that the polyelectrolytes contain free Na^+, and that the interaction between Na^+ and the polyelectrolytes is electrostatic in nature, and does not limit significantly the freedom of motion of the ion.

Selected Readings

1. Bloembergen N, Purcell EM, Pound RV. Relaxation Effects in Nuclear Magnetic Resonance Absorption. *Physical Review* 1948; 73(7):679–712.
2. Sprawls P, Bronskill MJ. *The Physics of MRI: 1992 AAPM Summer School Proceedings*. American Association of Physicists in Medicine, 1993, Woodbury, NY.

4 Fundamentals of Magnetic Resonance II: Imaging

4.1 Introduction

It has already been discussed that the phenomenon of magnetic resonance imaging (MRI) emerged from nuclear magnetic resonance (NMR) spectroscopy. Spectral acquisitions observed the spatio-temporal volume response of the nuclear resonance phenomenon from an ensemble of spins in a sample of interest. The fact that spins, and correspondingly the net magnetization vector $M(t)$, precess at an angular frequency $\omega_o = -\gamma.B_o$ led to the groundbreaking realization that introduction of specialized hardware in the spectrometers, known as *magnetic field gradients*, would allow spatial localization and encoding of information in the three dimensions of the Cartesian coordinate system. Such work was pioneered by Paul Lauterbur (1973; 1982), who won the 2003 Nobel Prize in Physiology or Medicine for the introduction of slice selection gradients for imaging. His groundbreaking work subsequently led to the publication of the first MRI of a human (Andrew et al. 1977; Hinshaw et al. 1979).

4.2 Magnetic Field Gradients

Linearly varying magnetic field gradients were incorporated to allow scanners to selectively excite particular anatomical slices of interest (slice selection), and encode the position of the resonating spins (or equivalently the net magnetization vectors from corresponding voxels of an image) using the resonant frequency-spatial position relationship (Figure 4.1).

The magnitude of the magnetic field changes linearly in the direction of the gradient. Each gradient coil is controlled independently and can be pulsed with current to generate a small, local field gradient that imposes a corresponding spatially and linearly varying angular frequency change on the spins (the larger the spin distance from the isocenter, the larger its precessional frequency).

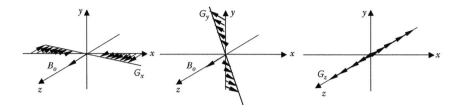

Figure 4.1. Magnetic field gradients G_x, G_y, and G_z and their spatial variations.

(The instrumentation technology for such coils will be described in Chapter 7.)
Mathematically,

$$G_x = \frac{\partial B_z}{\partial x} \tag{4.1}$$

$$G_y = \frac{\partial B_z}{\partial y} \tag{4.2}$$

$$G_z = \frac{\partial B_z}{\partial z} \tag{4.3}$$

If r denotes the vector location of the spin or magnetization vector at position (x, y, z), the angular frequency of any spin (or magnetization vector) in space, $\omega(r)$, can be written as

$$\omega_o(r) = -\hat{\gamma}(B_o + G_r.r) \tag{4.4}$$

where

$$\hat{\gamma} = \frac{\gamma}{2\pi} \tag{4.5}$$

This equation is the crux of *spatial encoding* in magnetic resonance imaging. The magnitude of such a gradient is very small with respect to the external magnetic field. Typically, a current of approximately 100 Anterior is passed through each set of conductors for the three gradients. The maximum amplitude of the gradients for clinical imaging is of the order of 1–2 Gauss/cm, and the rise times of the order of 0.5–1 ms or less. Although from an engineering perspective, faster switching times and stronger gradients can be utilized, nevertheless, strict Food and Drug Administration (FDA) limitations and guidelines limit the maximum slew rate in clinical scanners to approximately 40 Tesla/s (thereby avoiding neural stimulation upon human exposure).

In contrast to conventional imaging, NMR microscopy employs fast and much more powerful gradient sets than those of a typical system, to achieve smaller field of views (FOVs) and finer spatial resolution (the voxel dimensions decrease by a factor of 100 and a resolution of about 10 μm can be achieved).

Remarkably, patients and observers are often surprised from the pulsation of the gradients and relevant noise. Such effects are the result of increased Lorentzian forces exerted on pairs of conducting wires of the gradient coils placed in close proximity, induced from the passage of large currents. Ultra-fast imaging techniques, such as echo planar imaging (EPI), utilize gradient sets that can generate very high slew rates, and ultra-small rise and fall times. The inception and development of ultra-fast techniques was pioneered by Sir Peter Mansfield, who shared the Nobel Prize in Medicine and Physiology with Paul Lauterbur in 2003.

4.3 Spin–Warp Imaging and Imaging Basics

For the sake of discussion and explanation of the fundamentals of the formation of MR images, reference is made to *spin–warp imaging*. Such a term was inherited in the MRI literature from the publication of Edelstein et al. (1980). The term *warping* was used to exemplify the fact that use of gradients "warps" the spins by modifying their precessional frequencies. In the formation of an image, an anatomical slice is selected by exciting spins in the plane of interest, and information encoding is applied along the other two orthogonal directions, x and y. Encoding along the two orthogonal directions is based on the same principles of spin physics and is achieved by pulsing independently orthogonal gradients, known as the *frequency* and *phase* encoding gradients. By convention, the slice selection is often denoted along the z direction, frequency encoding along the x direction, and phase encoding along the y direction. In some cases (including cardiac imaging), the phase and frequency directions can be exchanged to avoid artifacts and improve image quality. Therefore, reference is often made to the physical (G_x, G_y, G_z) and virtual axes ($G_{frequency}$, G_{phase}, G_{slice}) of the three gradients. In the sections that follow, the physical principles that underline slice, frequency, and phase encoding are described and are explained by endorsing a mathematical formulation.

4.4   Slice Selection

Slice selection or spatial localization is based on the fact that signal can only be detected if the magnetization vector $M(t)$ tips away from the z axis, that is, if it nutates such that a transverse component is created. Although the radio frequency (RF) coil can detect nonselectively the entire volume of tissue, selectivity is imparted by tipping the magnetization vector to the transverse plane only from the spins within the slice of interest. So the problem of achieving slice selection reduces to selecting spins within a particular plane of interest by tipping their magnetization vector away from the z axis.

In the absence of a magnetic field gradient, all spins within the sample resonate at the Larmor frequency. Upon irradiation of the sample with an RF pulse, the spins (or equivalently the net magnetization vector) nutate away from the longitudinal direction, and the transverse component of the vector elicits a signal from the entire sample, which is detected by the coil antenna. The type of such an RF pulse is known as a *nonselective* pulse. The basis of slice selection uses the fact that switching on a gradient introduces a *spatial dependence of spin frequency with position along a given direction r* (refer to Equation 4.4). Turning on a G_z gradient causes, for example, spins to resonate at progressively increasing frequencies along the $+z$ axis. The spread of spin frequencies is linear and covers the entire object. If an RF pulse is transmitted with a bandwidth such that nutation of the magnetization vectors of only a subset of spins is achieved (only for those spins within the slice or volume of interest) while the constant z gradient is on, then spatial localization is achieved. In fact, for an excitation pulse of short duration, which imparts a small tip angle, it can be shown that the Fourier transform of its pulse shape can determine the profile of the tip angles through the excited slice (Pauly et al. 2001). This is known as a slice-selective pulse.

Mathematically, the bandwidth of the applied pulse must be such that

$$BW = \Delta f = f_{max} - f_{min} = \left(\frac{\gamma}{2\pi}\right)[(B_o + G_z . z_{max})$$

$$- (B_o + G_z . z_{min})] = \left(\frac{\gamma}{2\pi}\right) . G_z . \Delta z = \left(\frac{\gamma}{2\pi}\right) . G_z . ST \tag{4.6}$$

where ST represents the slice thickness. The bandwidth of the pulse Δf is centered about the frequency corresponding to the resonant frequency of spins at the center of the excited slice (Figure 4.2). In general, the excitation profile of the applied RF pulse is given by the Fourier transformation of its time domain shape (Figure 4.3). So for a rectangular pulse of bandwidth Δf,

$$S_{rf}(f) = rect\left(\frac{f}{\Delta f}\right) = \begin{cases} 1 & if \quad |f| < \Delta f \\ 0 & otherwise \end{cases} \tag{4.7}$$

Therefore, for a sinc pulse $S_{rf}(t) = A.f. \sin(\pi ft)/(\pi ft)$ with a 2 ms pulse width (distance between the first zero crossings), the bandwidth can be calculated as 1 kHz. Equivalently, at 64 MHz, for a pulse with a 1 kHz bandwidth, and a slice selection gradient of 1 Gauss/cm, the slice thickness is approximately 2.5 mm.

Important also is the consideration that such approximations apply in the cases of small nutation angles, and short RF pulse durations, in the presence of constant slice selection gradients where the Fourier transform approximation of the RF pulse shape-slice profile applies. The interested reader is referred to the paper by John Pauly et al. (2001) on a k-space analysis of small tip angle excitation.

An implication of slice selection is the fact that at the end of the applied gradient, G_z, the magnetization vectors exhibit a linear phase shift in space. In simpler terms, the spins within the excited slice resonate at different frequencies depending on their position. Before further warping is imposed on the spins, such a linear phase shift needs to be nulled. For this purpose, a *refocusing gradient,* that

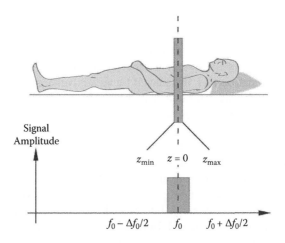

Figure 4.2. Slice selection and slice-selective RF pulse applied along the z direction.

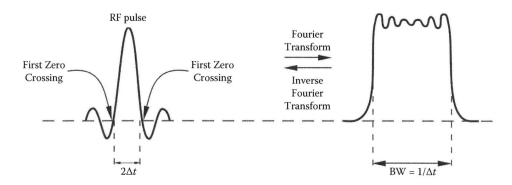

Figure 4.3. Left: Temporal variation of the slice-selective sinc RF pulse. Right: The corresponding slice-selective profile with an excitation gradient of 1 Gauss/cm. (Adapted from a figure by Professor E. McVeigh. With permission.)

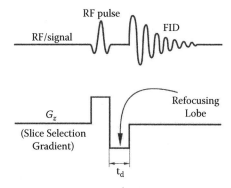

Figure 4.4. Application of a refocusing pulse after slice selection. If no other gradients are switched on, a free induction decay (FID) is observed immediately after the refocusing lobe. (Adapted from a figure by Professor E. McVeigh. With permission.)

is, a gradient with equal duration and equal and opposite amplitude to the slice selecting gradient, is applied (Figure 4.4).

4.5 Multislice and Oblique Excitations

The principle of multislice excitation is based on off-center slice excitation. In simple terms, spin frequencies that are offset from the center resonant frequency Δf_{off} are excited by modulation of the applied RF pulse so that its bandwidth is centered at the center frequency of the excited slice. In the time domain, this is equivalent to modulating the slice-selective pulse according to

$$s_{multislice}(t) = s(t).e^{-i2\pi\Delta\, f_{off}t} \tag{4.8}$$

Reference was made in an earlier section on the relationship of physical and virtual gradients, that is, the gradients for slice (G_{slice}), frequency (G_{freq}), and phase encoding (G_{phase}). When an arbitrary slice orientation is to be excited, or the direction of the frequency, phase, and slice-selective gradients is to be changed (elimination of artifacts, etc.), this can be achieved by modification of the so-called rotation matrix, that is, the matrix that relates the physical and virtual gradients. The rotation matrix (for an axial slice) is defined as

$$
\begin{bmatrix} G_x \\ G_y \\ G_z \end{bmatrix} = \begin{bmatrix} 1 & 0 & 0 \\ 0 & 1 & 0 \\ 0 & 0 & 1 \end{bmatrix} \cdot \begin{bmatrix} G_{freq} \\ G_{phase} \\ G_{slice} \end{bmatrix}
\tag{4.9}
$$

Generalization of the matrix implies selection of appropriate entries of the matrix elements to match the equation coefficients that determine the mathematical equation of any plane (including oblique slice selections).

Example 4.1

a. Design a spin–echo pulse sequence with which you could excite and image multiple slices Δz, in a fashion similar to the three axial slices shown in Figure Ex4.1.

b. If the thickness of each slice is 1 mm and the slice selection gradient amplitude $G_z = 1$ Gauss/cm, how much is the RF excitation pulse duration? Assume that the excitation pulse is defined by sinc = $\sin(x)/x$.

c. If the frequency encoding gradient $G_x = 1$ Gauss/cm, what is the field of view (FOV$_x$)?

d. If the constructed image consists of 256 × 256 pixels, what is the spatial resolution along the x axis?

ANSWER

a. Use the modulated sinc pulse for slice selection of the second, third, fourth, … tenth slices (Figure Ex4.2):

$$s(t) = \mathrm{sinc}(t) \tag{Ex 4.1}$$

$$s_{\mathrm{mod}}(t) = \sin c(t) e^{2\pi i \Delta f_{off} t} = s(t) e^{i\omega_{\mathrm{mod}} t} \tag{Ex 4.2}$$

where Δf_{offset} is determined by $\Delta = 5$ mm and BW$_z = \gamma G_z(ST)$. In accordance to Fourier transform and its properties:

$$FT[s(t)] = FT[\sin c(t)] = rect(\omega)$$
$$\tag{Ex 4.3}$$
$$FT[s_{\mathrm{mod}}(t)] = FT[s(t)] * FT[e^{i\omega_{\mathrm{mod}} t}]$$

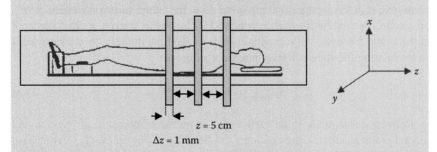

$z = 5$ cm

$\Delta z = 1$ mm

Figure Ex4.1. Multislice selection in axial orientation.

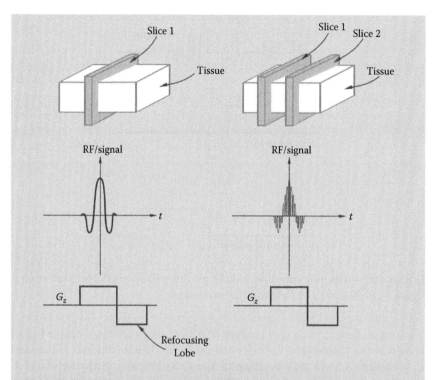

Figure Ex4.2. Single-slice and multislice excitation.

$$= rect(\omega).\delta(\omega - \omega_{mod}) = rect(\omega_{mod}) \qquad \text{(Ex 4.4)}$$

In effect, the frequency selection shifts in accordance to the modulating frequency ω_{mod}.

b. Given a sinc pulse excitation the frequency slice-selection profile is almost rectangular (according to the Fourier transform properties). Therefore:

$$\Delta t = \frac{1}{\gamma G_z ST} = 2\,\text{ms} \qquad \text{(Ex 4.5)}$$

which implies that the temporal width of the first zero crossings of the sinc pulse is 4 ms.

c. Assuming a standard analog-to-digital (A/D) converter readout sampling period of $T = 4\,\mu s$, then

$$\text{FOV}_x = \frac{1}{\gamma G_x T} = 58.7\,\text{cm} \qquad \text{(Ex 4.6)}$$

d. The pixel size is thus 58.7/256 = 2.3 mm.

4.6   Frequency Encoding

The physical mechanism of frequency encoding is similar to slice selection. A constant gradient is applied along the x axis that imparts a linear phase on

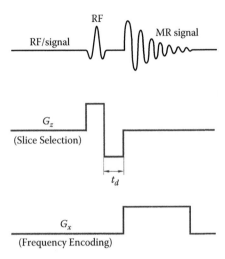

Figure 4.5. Frequency encoding and the resulting FID signal. (Adapted from a figure by Professor E. McVeigh. With permission.)

the excited spins, varying with position. The frequency encoding gradient is also known as the readout gradient since the analog-to-digital converter is immediately turned on, right after its onset, to sample and digitize the received signal. Frequency encoding is performed to allow mapping of the proton density (within each voxel of interest) to a position along one of the two spatial dimensions of the reconstructed image (Figure 4.5).

If we assume that the receiver coil that is used to detect the transverse magnetization has a uniform sensitivity throughout the volume of interest, that the sample is characterized by the relaxation values T_1, and T_2, and that the two-dimensional function that characterizes the proton density is $\rho(x, y)$, the received signal of the resulting FID can be expressed as

$$s_{fid}(t) = e^{-(t+t_d)/T_2} \int \int \rho(x,y)\, dx\, dy \tag{4.10}$$

where t_d is the time delay between the center of the applied slice-selective RF pulse and the onset of the frequency encoding gradient. The $e^{-(t+t_d)/T_2}$ term accounts for the T_2 relaxation that occurs after the application of the RF pulse. If the frequency encoding gradient G_x is turned on after a 90° slice-selective pulse, the equation of the resulting signal becomes

$$s(t) = e^{-(t+t_d)/T_2} \int \int \rho(x,y) \cdot e^{-i\gamma G_x xt}\, dx\, dy \tag{4.11}$$

For the imaging-adept readers, this equation is simply the Fourier transform of the projection of the proton density function $\rho(x, y)$ along the y direction on the x axis (recall the Fourier slice theorem in back-projection reconstruction in computer tomography). In simpler terms, the signal is the summation of column projections along the x axis for each pixel of the reconstructed image. Obviously, the task is to reconstruct the two-dimensional proton density function $\rho(x, y)$, that is, the distribution of the water molecules within the excited slice.

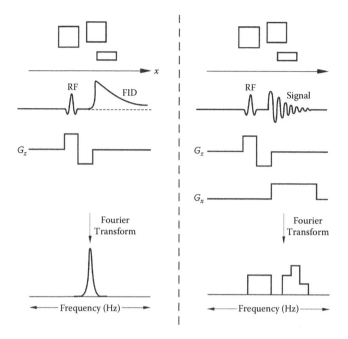

Figure 4.6. Frequency encoding in the absence (left) and presence (right) of a frequency encoding gradient. (Adapted from a figure by Professor E. McVeigh. With permission.)

Diagrammatically, the presence of the frequency encoding gradient and its effects is shown in Figure 4.6 as originally presented by McVeigh (1996).

4.6.1 Signal from a Point and Multiple Objects

It is also instructive to consider the mathematical formulation behind frequency encoding. This formulation can provide intuition into the reconstruction process, and will allow better understanding of advanced concepts and data acquisition schemes, described later on. The easiest approach is to consider the signal and its mathematical representation when it arises from a point object, followed by two and then n-point objects.

If the proton density function was composed of a single point object on the x axis at a position x_o, with intensity S_o, it can be written:

$$\rho(x, y) = S_0 . \delta(x - x_0, y) \tag{4.12}$$

$$s_o(t) = e^{-(t+t_d)/T_2} \int\int S_o . \delta(x - x_o, y) . e^{-i\gamma G_x x t} \, dx \, dy \tag{4.13}$$

$$s_o(t) = S_o . e^{-(t+t_d)/T_2} e^{-i\gamma G_x x_0 t} \tag{4.14}$$

The signal can be simply perceived as the signal of a decaying oscillator at position x_o. Measuring the frequency of the oscillator will provide its spatial position. Knowledge of the signal value also yields the true amplitude of the proton density content S_o. Extension of this formulation for n-point objects (Figure 4.7) yields

$$\rho(x, y) = S_0 . \delta(x - x_0, y) + S_1 . \delta(x - x_1, y) + ... + S_{n-1} . \delta(x - x_{n-1}, y) \tag{4.15}$$

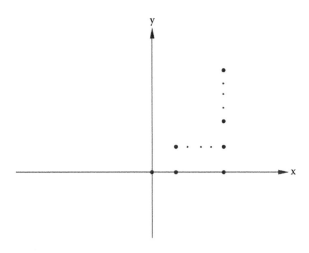

Figure 4.7. Mathematical formulation of frequency encoding based on single, two, and multiple point objects.

$$s_1(t) = e^{-(t+t_d)/T_2} \int \int [S_0.\delta(x-x_0,y) + S_1.\delta(x-x_1,y)].e^{-i\gamma G_x xt} \, dx \, dy \qquad (4.16)$$

.

.

.

$$s_{n-1}(t) = e^{-(t+t_d)/T_2} \int \int \begin{matrix} [S_0.\delta(x-x_0,y) + \\ S_1.\delta(x-x_1,y) + S_2.\delta(x-x_2,y) + ... \\ +S_{n-1}.\delta(x-x_{n-1},y)].e^{-i\gamma G_x xt} \end{matrix} dx \, dy \qquad (4.17)$$

So, to accomplish the task for solving and extracting the true amplitude values and special positions of the two- and n-point oscillators, there needs to be a solution to the two- and n-linear set of equations with 4 and $2n$ unknowns, respectively. A separate set of independent measurements is thus needed to solve such a system, available only with *phase encoding*.

4.7   Phase Encoding

In a similar fashion to frequency encoding, phase encoding imparts an additional phase to the spins (or increase in their respective precessional frequency, along the y direction) by turning on a separate gradient G_y for a time period t_y. If G_x and G_y are turned on simultaneously, then the data acquisition and image formation would resemble projection reconstruction used in computer tomography. Instead, the phase encoding gradient is turned on separately and independently from the frequency encoding gradient. It is pulsed in n steps. Following the signal formulation of the FID, consideration of phase encoding yields

$$s(t) = e^{-(t+t_d)} \int \int \rho(x,y).e^{-i\gamma(G_x xt + G_y t_y y)} \, dx \, dy \qquad (4.18)$$

4.7.1 Composite Signal from a Point and Multiple Objects

For a point object at x_o, y_o (Figure 4.7) the composite signal equation thus becomes

$$s_0(t) = e^{-(t+t_d)/T_2} \int\int S_0 . \delta(x - x_o, y - y_o) . e^{-i\gamma(G_x x_o t + G_y t_y y_o)} dx\, dy \qquad (4.19)$$

$$s_0(t) = e^{-(t+t_d)/T_2} S_0 . e^{-i\gamma(G_x x_o t + G_y t_y y_o)} = e^{-(t+t_d)/T_2} S_0 . e^{-i\gamma G_x x_o t} e^{-i\gamma G_y t_y y_o} \qquad (4.20)$$

which can be perceived as the multiplication of two decaying oscillators. If we consider the profile of this point object along the x direction, the above equation becomes

$$S(x) = e^{-(t+t_d)/T_2} . S . e^{-i\gamma G_y t_y y_o} . \delta(x - x_o) \qquad (4.21)$$

So, from the phase of the projection the position of the point object can be estimated. Extending this for two and n objects for m phase encoding steps yields a generalized set of projection equations:

$$S_n^m(x) = [S_{0,0} . e^{-i\gamma(G_0^0 t_y y_0)} + S_{0,1} . e^{-i\gamma(G_0^0 t_y y_1)} + S_{0,2} . e^{-i\gamma(G_0^0 t_y y_2)}$$

$$+ S_{0,m-1} . e^{-i\gamma(G_0^0 t_y y_{m-1})}] \delta(x - x_o) + \ldots$$

$$+ [S_{1,0} . e^{-i\gamma(G_1^1 t_y y_0)} + \ldots$$

$$+ S_{1,1} . e^{-i\gamma(G_1^1 t_y y_1)} + S_{1,2} . e^{-i\gamma(G_1^1 t_y y_2)} + \ldots \qquad (4.22)$$

$$+ S_{1,m-1} . e^{-i\gamma(G_1^1 t_y y_{m-1})}] \delta(x - x_1) +$$

$$\vdots$$

$$+ [S_{n-1,0} . e^{-i\gamma(G_{n-1}^{m-1} t_y y_0)} + S_{n-1,1} . e^{-i\gamma(G_{n-1}^{m-1} t_y y_1)}$$

$$+ S_{n-1,2} . e^{-i\gamma(G_{n-1}^{m-1} t_y y_2)} + \ldots + S_{n-1,m-1} . e^{-i\gamma(G_{n-1}^{m-1} t_y y_{m-1})}] \delta(x - x_{n-1})$$

$$S_n^m(x) = \sum_{k=0}^{n-1} \sum_{l=0}^{m-1} S_{k,l} . e^{-i\gamma G_k^l t_y y_l} \qquad (4.23)$$

To solve for the $n \times m$ equations with $2n \times m$ unknowns, the phase encoding gradient (G_y) needs to be stepped at multiple equal intervals such that

$$G_y^m = m . \Delta G_y \qquad (4.24)$$

This leads to the most generalized form of the *imaging equation* in discrete time (since the MRI signal is sampled as it evolves with time), in accordance to

$$s(n,m) = e^{-(nT+t_d)/T_2} \int\int \rho(x,y) . e^{-i\gamma(G_x x n T + m \Delta G_y y t_y)} dx\, dy \qquad (4.25)$$

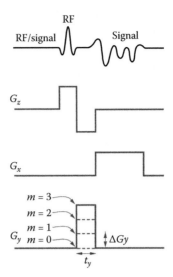

Figure 4.8. The timing diagram for spin–warp imaging. The phase encoding gradient is pulsed through multiple steps at a different and nonoverlapping time with respect to the frequency encoding. (Courtesy of Professor E. McVeigh.)

where n, m represent the discrete time indices and T represents the sampling period. This form of the imaging equation describes best spin–warp imaging, which can now be represented by the timing diagram of Figure 4.8.

4.8   Fourier Transformation and Image Reconstruction

A unique feature of MRI is the fact that the raw data are complex (unlike most other modalities that deal with real data). The data are the Fourier transform of the image rather than the image itself. Therefore, a number of important properties of the Fourier transformation can be used in generating the image and in data acquisition, including the complex symmetry of data in Fourier space and others.

Reconstruction of the image involves, in its simplest form, application of a fast 2D discrete Fourier transformation. Image reconstruction algorithms are an active and very interesting area of research in MRI. In Chapter 5 some unconventional data sampling trajectories (and image reconstruction algorithms) are presented.

Selected Readings

1. Carr HY, Purcell EM. Effects of Diffusion on Free Precession in Nuclear Magnetic Resonance Experiments. *Physical Review* 1954; 94(3):630–638.
2. Edelstein WA, Hutchison JMS, Johnson G, Redpath T. Spin–Warp NMR Imaging and Applications to Human Whole-Body Imaging. *Physics in Medicine and Biology* 1980; 25(4):751–756.
3. Lauterbur PC. NMR Zeugmatographic Imaging in Medicine. *Journal of Medical Systems* 1982; 6(6):591–597.

4. Morris PG. *Nuclear Magnetic Resonance Imaging in Medicine and Biology.* Oxford University Press, 1986, Oxford.
5. Stark D, Bradley W. *Magnetic Resonance Imaging.* Mosby, 1999, St. Louis, MO.
6. Sprawls P, Bronskill MJ. *The Physics of MRI: 1992 AAPM Summer School Proceedings.* American Association of Physicists in Medicine, 1993, Woodbury, NY.

5 Fundamentals of Magnetic Resonance III: The Formalism of *k*-Space

5.1   Introduction

In the previous chapter spin–warp imaging was introduced and discussed. It was shown how a constant gradient in the presence of a band-limited radio frequency pulse can be used to select a particular anatomical slice for excitation, and the received signal (after the readout gradient is turned on) becomes the projection of the object on the frequency encoding axis. It was proved mathematically that on each spatial location along the x axis, the composite signal is the summation of all the signal sources in columns perpendicular to that location. Mathematically, to be able to extract both the amplitudes and the locations of the signal sources along the orthogonal axis, a separate set of measurements needs to be carried out. Similar to frequency encoding, phase encoding is applied by imparting phase shifts on the spins along the y direction, depending on their position. Following simple linear algebra notation, the solution to decoding all amplitudes and spatial locations arises from pulsation of the phase encoding gradient at m steps, providing independent measurements that allow solution of the set of equations.

In this section, the formulation of frequency sampling is extended to its more general form and k-space sampling is introduced. Spin–warp imaging is used initially to describe the concepts, followed by a more general explanation for arbitrary trajectories and sampling schemes. In a natural progression, these concepts are extended to describe the notion of pulse sequences with a notable reference to an ultra-fast imaging acquisition technique known as echo planar imaging (EPI).

5.2   MRI Signal Formulation

The most general form of the imaging equation was shown in Chapter 4 to be

$$s(n,m) = e^{-(nT+t_d)/T_2} \int \int \rho(x,y) e^{-i\gamma(nTG_x x + m \Delta G_y y t_y)} \, dx \, dy \qquad (5.1)$$

$$0 \le n < N \qquad (5.2)$$

$$\frac{-M}{2} + 1 \le m \le \frac{M}{2} \qquad (5.3)$$

This is the representation of the time signal that is sampled at N intervals (with a period T) during frequency encoding, and that is phase encoded through m steps. So, for a 256×128 image matrix, 256 independent samples need to be collected, and 128 independent phase encoding steps need to be pulsed. Equivalently, this amounts to repeating the spin–warp imaging 256 times, one for each of the phase encoding steps of the G_y gradient.

It can be easily observed that the exponential arguments,

$$k_x = \gamma \cdot nT \cdot G_x \qquad (5.4)$$

$$k_y = \gamma \cdot m t_y \cdot \Delta G_y \qquad (5.5)$$

have dimensions of spatial frequencies. If we assume that the total sampling time and time delay between the center of the radio frequency (RF) pulse and the beginning of the readout gradient are small enough with respect to the transverse relaxation time T_2, the imaging equation becomes

$$s(k_x, k_y) = \int \int \rho(x,y) . e^{-i(k_x x + k_y y)} \, dx \, dy \qquad (5.6)$$

This equation states that the elicited time-sampled magnetic resonance imaging (MRI) signal is the Fourier transformation of the two-dimensional distribution of the proton density within the excited slice. Inverse discrete Fourier transformation of the time signal $s(k_x, k_y)$ yields the reconstructed image in the spatial dimensions $\rho(x, y)$. The notion of k-space (originally proposed by Truman Brown (1987)) is thus introduced in an analogy similar to Fourier space. k-Space can be simply thought of as the spatial frequency domain where sampling of the proton density function $\rho(x, y)$ occurs.

5.3 k-Space Formalism and Trajectories

Following the definition of phase accrual with time as the time integral of the angular frequency, k_x and k_y can be simply perceived as the spatial frequencies in the two dimensions x and y, having units of radians per meter. Generalizing the concept, this leads to

$$s[k_r(t)] = \int \rho(x,y) . e^{-i.2\pi.k_r.r} \, dr \qquad (5.7)$$

$$k_r(t) = \gamma \int_0^t G(r,t) \, dt \qquad (5.8)$$

where $k_r(t)$ is the spatial frequency vector and $G(r, t)$ the spatial gradient function.

Referring to the earlier chapter on Fourier transformations (Chapter 1), the *k*-space is the spatial frequency domain where sampling of the proton density function occurs. Although a continuum of spatial frequencies exists for every function $\rho(x, y)$, sampling is band-limited. The *k*-space formalism can be used to easily conceptualize and represent the sampling trajectory of the data acquisition scheme (often referred to as the timing diagram of the pulse sequence). In the specialized case of spin–warp imaging, *k*-space sampling can be easily thought to follow a lexicographic pattern of sampling as shown in Figure 5.1.

Traversing *k*-space can be achieved in a number of different ways, with the *k*-space sampling trajectory being characteristic of the pulse sequence used, and the way the gradients are switched on and off. There are numerous conventional and unconventional *k*-space trajectories employed nowadays for imaging, with characteristic examples shown in Figure 5.2. The reader must be able to predict the pulse sequence diagrams from the *k*-space trajectory, and vice versa.

Knowledge of the maximum spatial frequencies in the two dimensions $(k_{x,max}, k_{y,max})$ and the spatial sampling interval Δk_x, Δk_y allows calculation of the resolution and fields of view of the resulting image according to

$$\Delta x = \frac{2\pi}{k_{x,\mathrm{max}}} \tag{5.9}$$

$$\Delta y = \frac{2\pi}{k_{y,\mathrm{max}}} \tag{5.10}$$

and

$$\mathrm{FOV}_x = \frac{2\pi}{\Delta k_x} \tag{5.11}$$

$$\mathrm{FOV}_y = \frac{2\pi}{\Delta k_y} \tag{5.12}$$

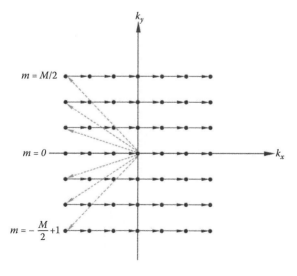

Figure 5.1. *k*-Space sampling using spin–warp imaging. The sampling trajectories follow a lexicographic order.

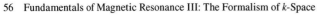

Figure 5.2. k-Space sampling along two unconventional trajectories. Left: Line sampling using an interleave acquisition scheme, where multiple k-space lines are acquired per view (in separate groups, that are subsequently interleaved to fill the k-space matrix). Right: Sampling along an Archimedean spiral trajectory.

Example 5.1

Plot the gradient waveforms $G_x(t)$, $G_y(t)$ that can achieve the following k-space trajectory (Figure Ex5.1).

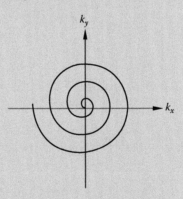

Figure Ex5.1. Spiral k-space trajectory.

ANSWER

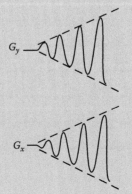

Figure solution Ex5.1: k-space trajectories for specified gradients.

5.4 Concept of Pulse Sequences

Discussion has been confined thus far to spin–warp imaging with a timing diagram representation shown in Figure 5.3.

Typically, reference to pulse sequences is associated with the timing diagrams of the RF chain, frequency, phase, and slice selection gradients. Often, the free induction decay (FID) or echo signal is indicated on the RF chain axis, together with the analog-to-digital conversion for data sampling. Since phase encoding is repeated by pulsing the G_y gradient m times, it is customary to depict the phase encoding with a lobe that is indicative of such process, thereby allowing its representation by drawing only the basic repeating part of the sequence, instead of the entire timing diagram. In Figure 5.3, drawn also are the echo (TE) and repetition times (TR), the times between the center of the RF pulse and the center of the echo, and to the center of the next RF pulse, respectively. Another interesting and important feature is the prephasor lobe with an area equal to half the area of the G_x gradient, along the readout gradient. This lobe accounts for and offsets the phase accrual on spins due to the frequency encoding gradient G_x. It ensures that there is no net phase accumulation on spins due to G_x at the middle of the echo.

Pulse sequences and pulse sequence design are a very active area of research in the field of MRI. A number of pulse sequence examples are discussed Chapter 6. Following next, however, is reference to an ultra-fast imaging technique known as echo planar imaging (EPI).

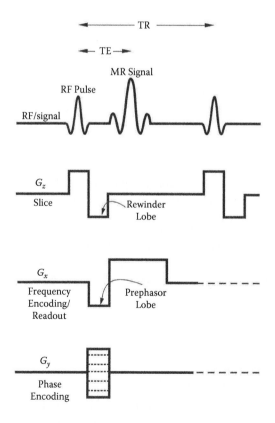

Figure 5.3. Spin–warp pulse sequence timing diagram. Marked are the echo (TE) and repetition times (TR).

Example 5.2

The use of gradient coils is fundamental for spin–warp imaging based on rectilinear Cartesian k-space sampling.

    a. Sketch a typical spin–warp pulse sequence diagram. Use symbols and letters to define the type of gradients used.

    b. How does spin–warp imaging relate to or differ from Paul Lauterbur's initial publication on zeugmatography?

    c. List the general imaging equation for spin–warp imaging.

ANSWER

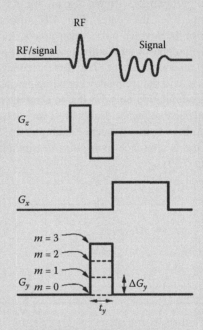

    a. See the above figure.

    b. Spin–warp imaging collects projections of signal-emanating decaying oscillators (lying in columns along the phase encoding direction) in the perpendicular direction known as the readout direction. Zeugmatographic imaging (in accordance to Paul Lauterbur's publication) is a generalized extension of this concept to arbitrary angles, a principle adopted in back-projection reconstruction in computer tomography.

    c. $s(n,m) = e^{\frac{(nT+t_y)}{T_2}} \iint \rho(x,y) e^{-i\gamma(G_x x nT + m\Delta G_y t_y y)}\,dx\,dy$ (Ex 5.1)

5.5   Echo Planar Imaging

Data acquisition schemes allow traversal and sampling of k-space in a number of different ways. Ultra-fast techniques can allow coverage and sampling of the entire k-space in as small as approximately 50–100 ms, the effective duration of the signal FID. Following the groundbreaking conception of EPI by Sir Peter

Mansfield and data acquisition after a single excitation, EPI was introduced for ultra-fast data collection and imaging. For his work, Sir Peter Mansfield shared the 2003 Nobel Prize in Physiology or Medicine with Professor Paul Lauterbur. Even if the inception of the technique occurred almost 27 years ago, only recently has such a technique found widespread use, due to the significant demands it places on the gradient hardware, the receivers, and the necessity for a highly uniform and homogeneous static magnetic field. In effect, the pulse sequence applies a single RF excitation and collects the entire k-space data immediately after.

Because it uses a single RF excitation, it has been alternatively known as a single-shot excitation. The term *blipped* is also used to describe the blip pulse in the phase encoding gradient that is applied to allow incrementing the k_y position in k-space.

Details relevant to the stringent hardware requirements, issues on the image quality, and artifacts will be addressed in the following chapters.

Selected Readings

1. Carr HY, Purcell EM. Effects of Diffusion on Free Precession in Nuclear Magnetic Resonance Experiments. *Physical Review* 1954; 94(3):630–638.
2. Cho ZH, Jones J, Singh M. *Foundations in Medical Imaging*. John Wiley & Sons, 1993, New York.
3. Edelstein WA, Hutchison JMS, Johnson G, Redpath T. Spin–Warp NMR Imaging and Applications to Human Whole-Body Imaging. *Physics in Medicine and Biology* 1980; 25(4):751–756.
4. Lauterbur PC. NMR Zeugmatographic Imaging in Medicine. *Journal of Medical Systems* 1982; 6(6):591–597.
5. Sprawls P, Bronskill MJ. *The Physics of MRI: 1992 AAPM Summer School Proceedings*. American Association of Physicists in Medicine, 1993, Woodbury, NY.
6. Stark D, Bradley W. *Magnetic Resonance Imaging*. Mosby, 1999, St. Louis, MO.
7. Morris PG. *Nuclear Magnetic Resonance Imaging in Medicine and Biology*. Oxford University Press, 1986, Oxford.
8. Mansfield P, Guilfoyle DN, Ordidge RJ, Coupland RE. Measurement of T1 by Echo-Planar Imaging and the Construction of Computer Generated Images. *Physics in Medicine and Biology* 1986; 31(2):113–124.

⑥ Pulse Sequences

6.1 Introduction

Pulse sequences (Cho 1993; Bernstein 2004) are the sets of time-spatial gradient functions that allow manipulation of the magnetization vector, sampling of k-space, and image formation. Different types allow different contrasts based on T_1, T_2, $T_2{}^*$, flow, diffusion, perfusion, and motion, with imaging times that range from milliseconds to seconds.

6.2 T_1, T_2, and Proton Density-Weighted Images

Common reference is often made to T_1-weighted, T_2-weighted, or proton density-weighted images. For any pulse sequence, T_1 weighting is achieved if TR ~ T_1 and TE ≪ T_2. Similarly, T_2 weighting is achieved when TR ≫ T_1 and TE ~ T_2, whereas proton density images are generated at fully relaxed conditions when TR ≫ T_1 (TR ~ 5 T_1) and TE ≪ T_2.

6.3 Saturation Recovery, Spin–Echo, Inversion Recovery

6.3.1 Saturation Recovery

In simple pulse-acquire experiments (Figure 2.6), an α pulse is used to tip the magnetization on the transverse plane. The recovery of the longitudinal component of the signal and the decay of the transverse magnetization vector are governed by

$$M_{SR}(t) = M_o e^{-\frac{TE}{T_2}} \left(1 - e^{-\frac{TR}{T_1}} \right) \tag{6.1}$$

6.3.2 Spin–Echo

Spin–echo (SE) was introduced by Hahn (1950). The basic timing sequence of gradient pulsation (pulse sequence) consists of two sequential (but spaced in time) radio frequency (RF) pulses. The first is 90°, followed by a 180° pulse. The first tips the

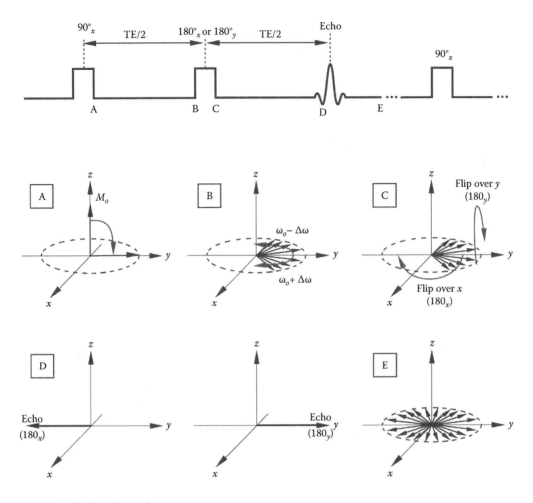

Figure 6.1. Spin–echo pulse sequence.

magnetization on the transverse plane, and the second pulse refocuses the dephased spins in the transverse plane, either on the +ve y axis or –ve y axis (Figure 6.1). The dephasing pattern of the magnetization vectors in the transverse plane is due to the presence of local field inhomogeneities that force the spins to "fall out of phase," with some spins rotating faster than others (with respect to the Larmor frequency).

If the inhomogeneities that cause the phase shifts on the rotating spins are constant in time, spin refocusing cancels/nulls these out. In the presence of inhomogeneities that vary temporally with time, there is magnetization loss due to incomplete rephasing. A sequence of 90–180° pulses in time is referred to as the Carr–Purcell–Meiboom–Gill (CPMG) sequence (Carr and Purcell 1954). These pulses lead to multiple echoes with maximum amplitude that falls exponentially with time, characterized by a time constant T_2, known as the transverse relaxation time (Figure 6.2).

Assuming ideal RF excitation pulses, single exponential decay behavior, and dispersal of transverse magnetization before the $(n + 1)$th repetition (Cho 1993),

$$M_{SE}(t) = M_o e^{-\frac{TE}{T_2}} \left(1 - 2e^{-\frac{[TR-TE/2]}{T_1}} + e^{-\frac{TR}{T_1}} \right) \qquad (6.2)$$

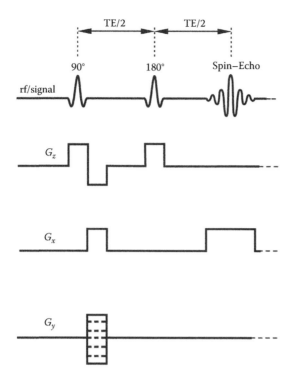

Figure 6.2. Two-dimensional spin–echo imaging.

If TE ≪ TR,

$$M_{SE}(t) = M_o e^{-\frac{TE}{T_2}} \left(1 - e^{-\frac{TR}{T_1}} \right) \tag{6.3}$$

6.3.3 Inversion Recovery

This sequence is designed to generate differential contrast in tissues with different T_1 relaxation behavior. The sequence inverts the longitudinal magnetization (from the $+z$ axis to the $-z$ axis) and, through the selection of TR = *TI*, nulls the magnetization from particular tissues of interest. The regrowth of the nulled magnetization and the relaxation of other tissues within the selected/excited slice tissue volume follows the T_1 recovery response of each structure.

The timing diagram of the infrared (IR) pulse sequence is shown in Figure 6.3 and differs from a conventional spin–echo due to the addition of the 180° RF pulse at the beginning of the sequence.

The first 180° selective pulse inverts the longitudinal magnetization, and the TI interval allows differential recovery based on tissue T_1. The slice-selective 90° tips the magnetization vectors in the transverse plane, followed by their dephasing due to local inhomogeneities in a manner similar to SE imaging. The echo forms at t = TE/2 after the second 180° refocusing pulse. The selection of TR is such that it allows complete recovery of the magnetization before the sequence is repeated. The inversion and TI relaxation widen the contrast differential between tissues based on their T_1 response. Thus, the sequence is considered a T_1-weighted sequence. Mathematically (Cho 1993),

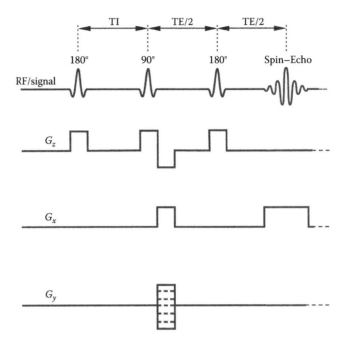

Figure 6.3. Inversion recovery imaging pulse sequence.

$$M_{IR}(t) = M_o \left(1 - 2e^{-\frac{TI}{T_1}} + 2e^{-\left(\frac{TR-TE/2}{T_1}\right)} - e^{-\frac{TR}{T_1}} \right) e^{-\frac{TE}{T_2}} \qquad (6.4)$$

But if TR $\gg$ TE,

$$M_{IR}(t) = M_o \left(1 - 2e^{-\frac{TI}{T_1}} + e^{-\frac{TR}{T_1}} \right) e^{-\frac{TE}{T_2}} \qquad (6.5)$$

In the case where TR $\gg T_1$,

$$M_{IR}(t) = M_o \left(1 - 2e^{-\frac{TI}{T_1}} \right) e^{-\frac{TE}{T_2}} \qquad (6.6)$$

To achieve nulling of the magnetization vector, *TI* must be chosen such that

$$TI = T_1 \ln 2 \qquad (6.7)$$

6.4   Gradient–Echo Imaging: FLASH, SSFP, and STEAM

6.4.1   Fast Low-Angle Shot (FLASH)

Gradient–echo (GRE) sequences were developed as fast to ultra-fast alternative techniques to spin–echo acquisitions. Unlike spin–echo techniques (multiecho approach after single RF excitation), Gradient–echo sequences are based on repetitive single RF excitation such as the sequence shown in Figure 6.4.

Conventional GRE sequences employ 90° RF excitation pulses. In fast low-angle shot (FLASH) imaging, however, small tip angles (about 10–15°) are used

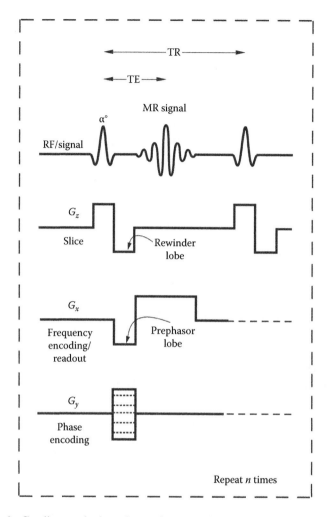

Figure 6.4. Gradient–echo imaging—the example of the FLASH pulse sequence.

that maintain the longitudinal magnetization almost unaffected and available for the subsequent excitation, TR ms later. Unlike spectroscopic imaging, where the nuclides exhibit long longitudinal relaxation times, where TR $\gg T_1$, in ^{1}H GRE sequences, TR can be really short.

The refocusing slice-selective gradient G_z restores the accrued phase of the spins along the z direction, and the phase encoding and readout gradients ensure adequate sampling and filling of k-space. The prephasor lobe in the readout gradient imparts negative phase on the spins such that refocusing occurs in the middle of the readout gradient. In such a way, the echo is recalled in the middle of the readout window.

For the general case of GRE imaging with flip angles $\alpha - 90°$, the imaging equation becomes

$$M_{GRE}(t) = M_o \frac{\left(1 - e^{-\frac{TR}{T_1}}\right)\sin\alpha}{1 + \cos\alpha \cdot e^{-\frac{TR}{T_1}}} e^{-\frac{TE}{T_2^*}} \tag{6.8}$$

Optimum signal performance is obtained when imaging is performed at $\alpha = \cos^{-1}[e - TR/T1]$, known as the Ernst angle, named after MRI Nobel Laureate Professor Richard Ernst (1987).

Example 6.1

The signal equation of ^{1}H MRI of a biological material employing the spoiled grass sequence is given by the following equation:

$$s(t) = e^{-\frac{TE}{T_2}} \frac{\left(1 - e^{-\frac{TR}{T_1}}\right)\sin\alpha}{1 - e^{-\frac{TR}{T_1}}\cos\alpha} \qquad (Ex\ 6.1)$$

where α is the tip angle, TR the time interval between the center points of successive RF excitation pulses, and TE the time interval between the center of the excitation pulse and the center of the echo. T_1 and T_2 are the relaxation parameters, respectively (longitudinal and transverse relaxation times).

Derive the equation of the tip angle that maximizes the signal intensity $s(t)$. How is this angle known? For your calculations assume that $TE \ll T_2$.

ANSWER

To compute the optimum tip angle a_{opt}, take the first derivative of the solution from the first part of the problem and set it to zero:

$$\frac{\delta S}{\delta a} = M_o \left(1 - e^{-\frac{TR}{T_1}}\right) e^{-\frac{TE}{T_2}} \frac{\delta}{\delta a}\left(\frac{\sin a_{opt}}{1 - \cos a \cdot e^{-\frac{TR}{T_1}}}\right) = 0 \quad (Ex\ 6.2)$$

and

$$\frac{\delta}{\delta a}\left(\frac{\sin a_{opt}}{1 - \cos a \cdot e^{-\frac{TR}{T_1}}}\right) = \frac{\left(\cos a_{opt} - \cos^2 a_{opt} e^{-\frac{TR}{T_1}}\right) - \left(e^{-\frac{TR}{T_1}}\sin^2 a_{opt}\right)}{\left(1 - \cos a \cdot e^{-\frac{TR}{T_1}}\right)^2} = 0$$

$$(Ex\ 6.3)$$

This means that

$$\left(\cos a_{opt} - \cos^2 a_{opt} e^{-\frac{TR}{T_1}}\right) - \left(e^{-\frac{TR}{T_1}}\sin^2 a_{opt}\right) = 0 \quad (Ex\ 6.4)$$

or equivalently:

$$\cos a_{opt} - e^{-\frac{TR}{T_1}}\left(\cos^2 a_{opt} + \sin^2 a_{opt}\right) = \cos a_{opt} - e^{-\frac{TR}{T_1}} = 0 \quad (Ex\ 6.5)$$

Therefore,

$$a_{opt} = \cos^{-1} e^{-\frac{TR}{T_1}} \qquad (Ex\ 6.6)$$

This is known as the Ernst angle.

6.4.1.1   Spoiling

A significant problem often encountered with gradient–echo (GRE) sequences is the generation of a spin–echo after the nth RF excitation due to the accumulation and rephrasing of the residual transverse magnetization. The artifact due to the spin–echo (in the phase encoding direction) can be eliminated by introducing a variable spoiling scheme (where a gradient, random in value, is executed, known as the spoiler gradient) in the direction of the slice selection (Figure 6.5).

The spoiling takes effect as an incremental gradient spoiler in a random fashion that eliminates the artifact by crushing any residual coherent magnetization. The spoiler gradient can be played out in any gradient axis direction as long as the effective area is large enough to avoid introduction of spurious artifacts in k-space.

A second, more efficient spoiling method is phase cycling. The phase of both the RF excitation and receiver is changed in a pseudorandom order to spoil the transverse magnetization at the end of each excitation. This method does not employ gradients and is thus power efficient.

Different acronyms have traditionally been used by different imaging companies for their implementation of pulse sequences. GRASS and spoiled GRASS are variants of FLASH and spoiled GRASS type sequences implemented by General Electric (GE). A listing of the pulse sequences and relevant acronyms from all imaging companies is shown in Table 6.1.

6.4.2   Steady-State Free Precession (SSFP)

In the cases where the transverse magnetization coherences are not nulled, dephased, or dispersed (often known as steady-state free precession), each RF pulse converts some of the longitudinal magnetization into transverse magnetization, and vice versa. The mathematics of the signal and its evolution become complicated. In this

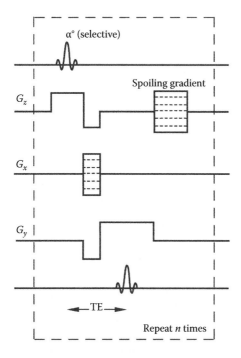

Figure 6.5. The concept of gradient spoiling.

Table 6.1. Major Pulse Sequences and Control Variables Used by Different MRI Companies

Pulse Sequence	Siemens	GE	Philips
Spin–echo	Spin–echo (SE)	Spin–echo	Spin–echo (SE)
Gradient echo	Gradient echo (GRE)	GRE, GRASS	Fast field echo (FFE)
Spoiled gradient echo	FLASH	SPGR	T_1FFE
Steady-state free process	TrueFISP	FIESTA	Balanced-FFE
Ultra-fast gradient echo	Turbo flash	Fast GRE, fast SPGR	Turbo field echo (TFE)
3D ultra fast	MP Rage	3DFGRE, 3D fast SPGR	3D TFE
Inversion recovery	IR, TurboIR, TIR	IR, MPIR	IR-TFE (or TSE)
Short tau IR	STIR	STIR	STIR
RARE sequence	Turbo spin–echo (TSE)	Fast spin–echo	Turbo spin–echo (TSE) GRASE
Rapid acquisition relaxation Enhancement	Turbo GSE, TGSE IPAT Phase contrast	GRASE ASSET Phase contrast SSFSE	SENSE PC velocity, phase contrast, quantitative
Gradient and spin–echo Flow velocity encoding Half Fourier single-shot turbo (Fast) spin–echo	HASTE		Flow Single-shot TSE
Cine study (cardiac)	CINE FLASH	Cine, FASTCARD	Cine
Spin–echo black blood (cardiac)	Dark-blood-prepared TSE, HASTE, FLASH	Double-IR FSE, FSE- XL with blood suppression	Black blood prepulse
Spin–echo black blood null fat (cardiac)	TRIM	Triple IR FSE, FSE-XL IR with blood suppression	Black blood prepulse plus SPIR
Viability imaging (cardiac)	Segmented 2D TurboFLASH	Myocardial delayed enhancement (MDE) IR prep-gated FGRE	Delayed hyperenhancement IR TFE
Myocardial perfusion (cardiac)	2D TurboFLASH SS TurboFisp sat or IR GRE-EPI	FGR-ET multiphase	Multislice perfusion, saturated TFE/EPI, B-FFE, TFE
Scan Parameters			
Echo time, repetition time	TE, TR in ms	TE, TR in ms	TE, TR in ms
Inversion time	TI in ms	TI in ms	TI in ms
Number of echoes per TR	Echo train length (ETL), turbo factor	ETL	TSE factor, turbo spin factor
Time between echoes	Echo spacing	Echo spacing	Echo spacing
Repeated measurements RF pulse in gradient echo	Acquisitions, number of averages	NEX Flip angle Acquisition time spacing	NSA Flip angle (FA) Acquisition time
Scan measurement time	Flip angle	Off-center FOV	Slice gap
Spacing between slices	Acquisition time (TA)		Off center (AP, LR, FH)
Shifting slices off center	Distance factor, In% Off-center shift		
Field of view	FOV	FOV	FOV
Nonsquare field of view	Rec FOV	Rectangular FOV PFOV	RFOV
Bandwidth		Receive bandwidth	Fat-water shift (opposite of BW)
Variable bandwidth	Optimized bandwidth	VB	Optimized bandwidth
Oversampling in frequency	Oversampling	Always on	Always on
Oversampling in phase	Phase oversampling	No phase wrap	Fold-over suppression
Segmented *k*-space	Lines, segments	Views per segment	Views, segments
Time delay/block *k*-space	Time delay (TD)	Intersegment delay	TD
Patient orientation scan	Localizer, scout	Localizer	Plan scan
Half Fourier imaging	Half Fourier	½ Nex, fractional NEX	Half scan, HS
Partial echo	Asymmetric echo	Fractional echo	Partial echo
Gradient moment nulling	GMR	Flaw Comp	Flow Comp, FC, flag

Continued

Table 6.1. (*Continued*) Major Pulse Sequences and Control Variables Used by Different MRI Companies

Pulse Sequence	Siemens	GE	Philips
Prep pulse—chemically	FAT SAT	FAT SAT, CHEM SAT	SPIR
Prep pulse—spatially	Presat	SAT	REST
Moving sat pulse	Travel sat	Walking sat	Travel REST
Image sync. with ECG	ECG triggered	Cardiac gated,	ECG triggered, VCG
Delay after *R*-wave	Trigger delay (TD)	triggering,	Trigger delay (TD),
Respiratory gating	Respiratory gated	trigger delay (TD)	trigger, PEAR
		Respiratory comp (RC),	
		respiratory triggered	

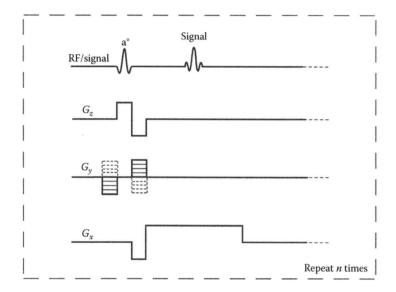

Figure 6.6. Steady-state free precession pulse sequence.

sense, and given the ultra-small TRs employed in this case, imaging occurs when the transverse magnetization reaches steady state (Figure 6.6).

Steady state is achieved relatively fast due to the ultra-small TR employed. The path to steady state ultimately depends on whether the sign of the RF pulse alternates (or not) from one TR to the next. Large flip angles are used (around 60°) to drive the magnetization to steady state. In practice, "dummy" pulses are usually played out at the beginning of the pulse sequence acquisition to attain steady state, a condition that the user can control via programmed pulse control variables (cv).

6.4.3   Stimulated Echoes (STEAM)

A stimulated echo can be obtained in direct extension to the spin–echo sequence considered earlier (Figure 6.7).

The sequence consists of three 90° pulses, all of which could be selective (or a sequence where the first two are nonselective, followed by a nonselective (or a selective) pulse). The first pulse tips the magnetization vector onto the transverse plane, allowing it to diphase over τ = TE/2. During the mixing time (TM) interval τ = *TM*, the magnetization experiences only T_1 relaxation (a mechanism that inherently allows control of T_1 contrast), followed by a slice-selective pulse and imaging. The stimulated echo forms at τ = TE/2 from the third RF pulse.

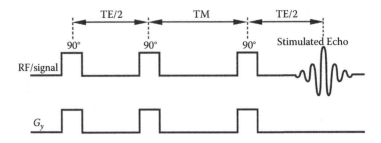

Figure 6.7. Stimulated echo pulse sequence diagram.

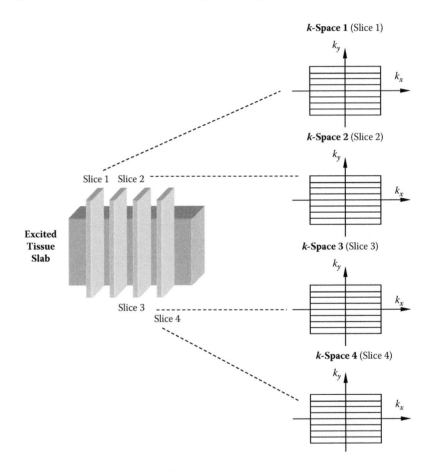

Figure 6.8. Principle of multislice imaging.

6.4.4   Multislice Imaging

Using any of the aforementioned techniques for data acquisition and imaging, single or multiple slices can be collected within the same time interval. This follows the realization that during the acquisition time (TR = TE), additional encoding can be performed to acquire and fill the k-space of adjacent slices in space. To achieve this, a modulated RF pulse is played out at a frequency that yields the appropriate phase/spectral offset. To avoid interference from residual magnetization in adjacent slices, excitation follows a spatially alternate slice pattern (e.g., excitation of slice 1 followed by excitation of slice 4, then slices 2 and 3) (Figure 6.8).

6.4.5 Volume Imaging

Contrary to 2D, single-slice, or multislice imaging, three-dimensional (3D) volume imaging requires excitation of spins within a volume of tissue. This can be achieved in a number of ways. There are two preferred methodologies; the first is an extension of spin–warp imaging (Figure 4.8) to 3D with minor alterations in the pulse sequence employed. Specifically, the differences with conventional single-slice spin–warp imaging are that the entire tissue slab (under consideration) is nonselectively excited, followed by the concurrent phase encoding along the y and z axes. Frequency encoding is conducted in a similar fashion to 2D spin–warp imaging, as shown in Figure 6.9.

The elicited MRI signal is governed by the imaging equation (Equation 4.25), modified to accommodate the referenced changes as follows:

$$s(n,m) = e^{-(nT+t_d)/T_2} \int \int \rho(x,y) \cdot e^{-i\gamma(G_x xnT + m\Delta G_y yt_y + n\Delta G_z zt_y)} \, dx\,dy\,dz \qquad (6.9)$$

Despite the simplicity of implementation of this technique for volume imaging, it is nevertheless associated with signal-to-noise ratio (SNR) inefficiency and subsequent image plane reconstruction complexities. To be able to reconstruct images along any oblique plane, high-resolution, isotropic imaging acquisitions must be completed a priori, a typically time-consuming and inefficient acquisition scheme (if lexicographic/Cartesian sampling is employed).

Alternative techniques are associated with unconventional (but more efficient) data acquisition schemes. The most notable examples are spiral or twisted projection imaging (Irarrazabal and Nishimura 1995; Boada et al. 1997; Noll 1997). In such schemes, a nonselective excitation is followed by 3D k-space coverage through simultaneous pulsation of the x, y, and z gradients. Sampling covers the k-space either along spiral trajectories (in an interleaved fashion) or along conical surfaces (or various diameters) (Figure 6.10). Twisted projection sequences are sometimes preferred since they avoid increased gradient amplitudes (an issue encountered with spiral imaging at later encoding times), and overcome hardware imperfections at the onset of gradients at initial sampling periods.

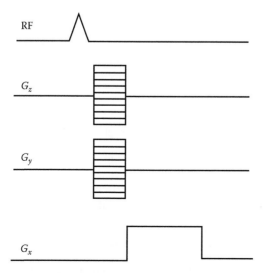

Figure 6.9. Spin–warp pulse sequence for (3D) volume imaging.

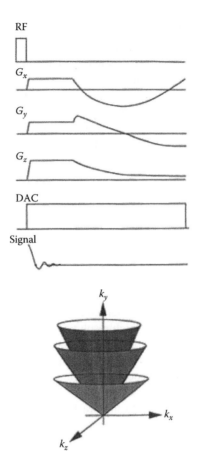

Figure 6.10. Top: A typical example of a twisted projection pulse sequence. Bottom: Corresponding conical k-space sampling surfaces.

Such schemes are highly efficient and can be optimized for 3D coverage of k-space. They are, however, associated with convoluted postprocessing and reconstruction. Data sampling at non-Cartesian loci are implicitly associated with the necessity for data regridding to the nearest neighbor Cartesian loci using weighting functions (depending on the distance of the actual, acquired k-space sample from its nearest rectilinear locus), followed by filtering and inverse 3D–FT (Boada et al. 1997).

6.5   Bloch Equation Formulation and Simulations

Although in most commonly used pulse sequences, closed-form expressions for the signal are available, nevertheless a more accurate approach is the complete simulation of the evolution of the magnetization vector, based on the Bloch equations. Bloch equation simulators are often available.

6.6   Technical Limits and Safety

Pulse sequences and gradient switching often impose stringent requirements on the hardware and software of the system in terms of power, efficiency, temperature, heating, eddy currents, etc. Quantitative assessment of these effects is often performed, and upper limits are compared, based on Food and Drug Administration (FDA) guidelines.

Selected Readings

1. Cho ZH, Jones J, Singh M. *Foundations in Medical Imaging.* John Wiley & Sons, 1993, New York.
2. Sprawls P, Bronskill MJ. *The Physics of MRI: 1992 AAPM Summer School Proceedings.* American Association of Physicists in Medicine, 1993, Woodbury, NY.
3. Morris PG. *Nuclear Magnetic Resonance Imaging in Medicine and Biology.* Oxford University Press, 1986, Oxford.
4. Bernstein MA, King KF, Zhou XJ. *Handbook of MRI Pulse Sequences.* Elsevier, 2004, Amsterdam.

7 Introduction to Instrumentation

7.1 Introduction

This chapter discusses basic instrumentation aspects as they relate to the magnetic resonance imaging (MRI) scanner and its functionality. These include the magnet and the various designs, matters on field homogeneity, stability, shimming, and the fringe field, gradient coils and radio frequency (RF) coils and their designs, other electronics (transmitters, receivers), and decoupling schemes. A general overview of the scanner and its auxiliary systems is shown in Figure 7.1.

Specifically, and in a block diagram form, the MRI system and the RF chains (transmitter, receiver, modulator, and receiver demodulator) are depicted in Figure 7.2. In particular, the transmitter-receiver, the transmitter modulator, and the receiver demodulator are drawn in Figure 7.3 in a more detailed block diagram form. Once the signal is received and demodulated to a frequency of a few kHz, a double-balanced demodulator is used and the signal is detected in quadrature. It is then sampled and inverse Fourier transformed for reconstruction (Figure 7.4).

7.2 Magnets and Designs

One of the most important criteria for superior performance and image quality is a highly uniform magnetic field over the imaging volume (typically of the order of 40–50 cm^3). To allow such homogeneity, while providing at the same time easy patient access, comfort of the patient, and minimal claustrophobia, magnets have endorsed designs of a cylindrical shape with a hollow cylindrical bore. Modern open and portable MRI systems have recently become available; however, they still lack somewhat in homogeneity performance with respect to closed-magnet (open-bore) systems (Figure 7.5).

The magnetic field function that is produced along the direction of the z axis can be expressed in terms of a Taylor series expansion as a summation of the first

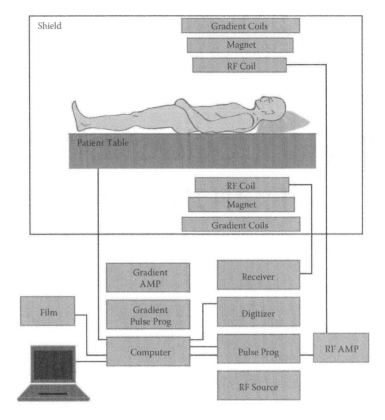

Figure 7.1. A schematic representation of all MR hardware and interconnections.

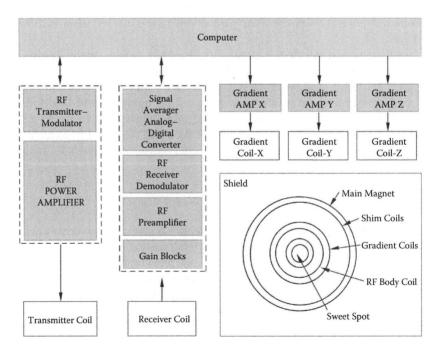

Figure 7.2. A schematic representation of the transmitter and receiver RF chains, gradient coils, and main magnet room.

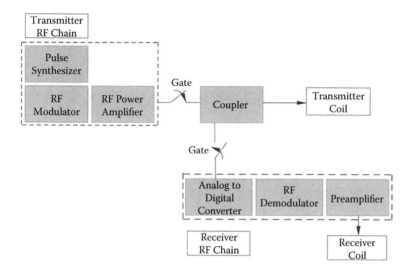

Figure 7.3. A schematic representation of the transmitter and receiver RF chains. Gates are used to isolate the receiver from the transmitter (during the receive phase and vice versa). Such isolation reduces coupling and noise.

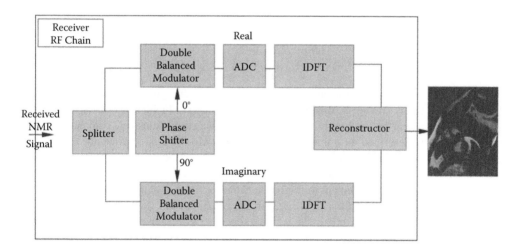

Figure 7.4. A schematic representation of the receiver RF chain and quadrature reconstructor.

field harmonic and higher-order terms that involve Legendre polynomials of degree n. In general the magnetic field at the center of the Cartesian axes (at $z = 0$) can be written as (Bottomley 2003)

$$B_z = J.L.R_o.F(\chi_1, \chi_2, ..., \chi_n) \tag{7.1}$$

where J is the current density in the coil conductor (current per unit area), L is the ratio of conductor cross-sectional area to the total area, R_o is the inner radius of one of the magnet coils, and F is a filling factor function that depends on a number of geometrical variables χ_i. The function F can be expanded

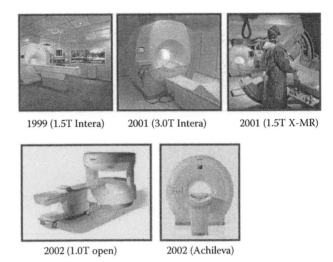

1999 (1.5T Intera)   2001 (3.0T Intera)   2001 (1.5T X-MR)

2002 (1.0T open)   2002 (Achileva)

Figure 7.5. Magnet systems and designs and their evolution since 1999 (Philips Medical Systems). (Reproduced from GyroTools, Inc., Lecture Notes, ETH, Zurich. With permission.)

using a Taylor series approximation of about $z = 0$, and the axial field can be written as

$$B_z = JLR_o \left[1 + e_2 \left(\frac{z}{R_o} \right)^2 + e_4 \left(\frac{z}{R_o} \right)^4 + \dots \right] \qquad (7.2)$$

where the coefficients e_n are dimensionless and geometry dependent. The odd terms in the expansion vanish by symmetry. Field homogeneity is optimum when the maximum number of low-order coefficients, e_n, vanishes. A magnet is described as a magnet with an order ith if the e_i coefficient vanishes. So, a Helmholtz pair is fourth order if e_4 vanishes, or equivalently, the expansion of B_z has terms up to and including e_3. Most commercial scanners are of order 6 or 8, and the requirement for higher field homogeneity is achieved with the use of additional shim coils (Figure 7.2) that through current flow adjustments (which are computer controlled) can achieve superior performance and field homogeneity.

7.2.1   Resistive Electromagnets

Most systems are built around a superconducting magnet or an electromagnet. Both types use electrical current to generate a uniform field. An adequate field is produced by several pairs of large-diameter coils of wire known as Helmholtz coils. The primary field is made as homogeneous as possible by moving about shim plates of steel (passive shimming) and by adjusting the small currents in various additional shim coils, which produce slight field corrections (active shimming).

So for an electromagnet, copper or aluminum conductors are wound on an aluminum frame with a bore diameter of about 1 m within which the gradient field coils and RF coils are fitted (Figure 7.6). An inner bore of about 40–60 cm is left for the patient. Conductor Joule heating occurs so the current-carrying conductors are hollow, to allow circulating cooling water to cool them down. Resistive magnets can attain field strengths of about 0.01–0.2 T.

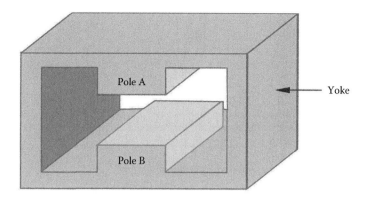

Figure 7.6. An iron core resistive magnet. The iron yoke provides a path for the return flux, and the pole faces and their geometry determine the field and its homogeneity.

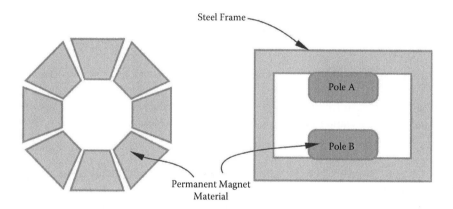

Figure 7.7. Types of permanent magnets.

7.2.2 Permanent Magnets

Large blocks of ferrous metal are used to generate the magnetic field. Unlike the other types, the main field of the permanent magnet is aligned vertically. Such magnets are largely maintenance-free and consume no electric power or cryogens (i.e., liquid helium and nitrogen), but their disadvantage is that they are capable of producing only relatively low fields (up to approximately 0.4 T).

Two basic designs, the ring dipole and the H-frame, are shown in Figure 7.7. Eight segments compose the ring dipole, with the magnetization indicated by the arrows. The H-frame consists of two massive pole pieces held together by a steel frame. Although bulkier in size and structure, the H-frame design is preferred due to improved fringe field advantages.

7.2.3 Superconducting

Most modern machines use a superconductor (Figure 7.8). A superconducting magnet exploits the significant reduction of electrical resistance in some materials at very low temperatures (e.g., niobium-titanium alloy at –253°C). Once the current is initiated, there is no further need for power input, nor are there any conducting losses.

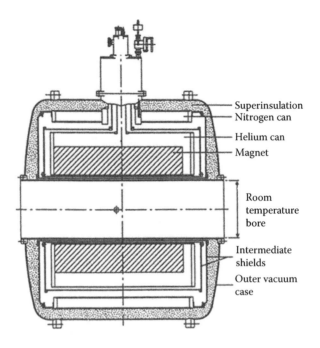

Figure 7.8. Left: A cross-sectional schematic of a superconducting magnet. (Reproduced from Thomas, S. R., Magnets and Gradients Coils: Types and Characteristics, in *The Physics of MRI: 1992 AAPM Summer School Proceedings*, American Association of Physicists in Medicine, Woodbury, New York, 1993. With permission.)

To maintain the superconducting condition, the entire coil must be immersed in liquid helium (approximately –269°C) contained within a cryostat (helium Dewar vessel), i.e., a stainless steel vacuum insulated cryostat that is filled with liquid helium. A surrounding refrigerator prevents warming of helium from the outside sources. Some systems use liquid nitrogen instead of a refrigerator system (boiling point of approximately –196°C) to prevent heat inflow to the helium container. Helium boils off slowly, so it has to be replenished every few months, and so does liquid nitrogen (even more frequently). The high cost associated with the MRI unit is mostly due to the manufacture of the magnet-cryogen assembly and its cryogenic power consumption (Figure 7.8).

In the United States, the Food and Drug Administration (FDA), currently limits clinical diagnostic studies on humans in magnets with a field strength of at most 3 T (i.e., 30,000 Gauss).

Overall, each type of a magnet has its own advantages and disadvantages, summarized in Table 7.1.

7.3   Stability, Homogeneity, and Fringe Field

A number of important features of the magnetic field and the magnet type itself, including the temporal field stability, field homogeneity, and fringe field, are critical for the image spatial resolution, and ultimate signal-to-noise ratio (SNR) performance. Temporal stability is determined by the magnet type and design and by external time-varying fields and interactions. Resistive magnets, for example, are highly dependent on power variations. Temperature fluctuations are also

Table 7.1. Comparison Chart of the Advantages and Disadvantages of Different Magnet Types

Magnet	Advantages	Disadvantages
Resistive	Low cost	Limited field uniformity
	No need for cryogens	Low maximum field strength (<0.5 T)
	Can turn it on and off	Dissipation of heat using water cooling system
		Medium-range fringe field
		Restricted patient access
Permanent	Low cost	Limited uniformity
	Zero power consumption	Low maximum field strength (<0.3 T)
	No need for cryogens	Sensitivity to temperature
	Confined fringe field	Bulky
	No heat generated	
	Good patient access	
Superconducting	Can achieve high field strengths	Cryogens required
	Increased field uniformity	Large spatial extent of fringe field
	Increased temporal stability	Restricted patient access (open magnets excluded, although associated with lower field homogeneity)
		Costly

extremely important for permanent magnets. Superconducting magnets are the most stable, with an expected resonant frequency shift at 1.5 T of less than 1 MHz in approximately 20 years.

Homogeneity is also a critical determinant of the image spatial resolution. It is usually defined in parts per million according to the ratio of the maximum static field (B_o) change, defined by

$$\frac{\Delta B_o}{B_o} = \frac{(B_{\max} - B_{\min})}{B_o} \qquad (7.3)$$

For clinical systems, field homogeneity is expected to be in the range of approximately 15 ppm within the central 40 cm diameter spherical volume. Within the "sweet spot" the FWHM of a line-width must be less than about 0.3–1.0 ppm.

Magnetic shielding is often used to reduce the fringe field, that is, the field extending outside the magnet room. Active or passive shielding is often achieved with specialized coils generating a field to counteract the fringe field extent, or alternatively with a ferromagnetic passive shield enclosure.

7.4   Gradient Coils

Following one of the introductory chapters, gradient coils (Figure 7.9) are used to generate the linear, spatially varying weak magnetic fields necessary for spatial encoding. Their design is based on use of multiple current-carrying conducting loops, similar to the designs of the main magnet types. A number of techniques are used to optimize the design of such coils, based on inductance, power dissipation, and field linearity. Below two characteristic examples are discussed.

Both the gradient coils and the radio frequency coils (as will be discussed in Sections 7.4.1–7.4.3 and 7.5) can use the Biot–Savart law to compute the magnetic

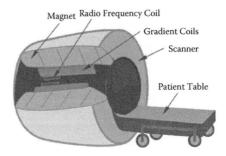

Figure 7.9. Schematic representation of a typical clinical scanner system.

field distribution. For a magnetic field in three-dimensional space $B(r)$, generated by a unit length of a conductor δl, carrying a current I, Biot–Savart states that for any point in space $P(x, y, z)$ the magnetic field is given by

$$B(r) = \frac{\mu_o I}{4\pi} \oint \frac{dl \times (r - r')}{|r - r'|^3} dr \qquad (7.4)$$

where μ_o is the magnetic permeability and r and r' are the distances of the conducting wire and the point in space for which the magnetic field is being computed, respectively.

In modern clinical scanners, gradient windings are wrapped around a cylindrical former in patterns that optimize linearity and minimize power consumption. Typical examples of axial gradient coils include the Maxwell pair and the Golay coils.

7.4.1   Maxwell Pair and Golay Coils

The Maxwell pair consists of a pair of current-carrying loops (of radius a), concentrically placed at a distance z apart, as shown in Figure 7.10. Considering that the two coil loops carry current in opposite polarities, the field along the z axis can be computed using the Biot–Savart law to be

$$B_z = \frac{\mu_o I a^2}{2[(z + L/2)^2 + a^2]^{3/2}} - \frac{\mu_o I a^2}{2[(z - L/2)^2 + a^2]^{3/2}} \qquad (7.5)$$

where L is the coil separation, a is the coil radius, I is the current, and μ_o is the magnetic permeability of the medium in which these are placed. Changing L changes the uniformity of the coil. Optimal uniformity is generated when $L = a\sqrt{3}$.

Transverse gradients (gradients that vary linearly in the x and y planes) are based on the same principle. Professor X. Golay was the first to introduce such coils, but the full analysis of spherical harmonics for calculation of the gradient fields in the transverse plane is beyond the scope of this book.

No field is produced in the z direction; the gradient is formed by the current arcs. There is a special configuration and geometry for nulling the higher-order harmonics, achieved by choosing the equatorial and azimuthal angles φ and ψ appropriately. The configuration in Figure 7.11 produces the G_y gradient. Rotating the same set of coils by $90°$ produces the G_x gradient.

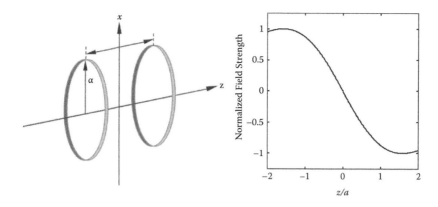

Figure 7.10. Left: The Maxwell pair—an example of an axial gradient coil. Right: Variation of the gradient field due to the two coils (each having a radius a) along the z axis.

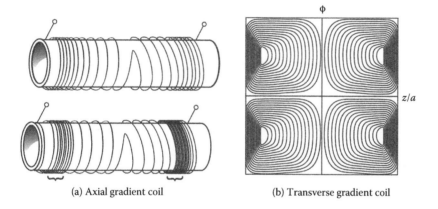

(a) Axial gradient coil   (b) Transverse gradient coil

Figure 7.11. Example of gradient winding schemes for an axial and a transverse gradient coil. (Reproduced from Thomas, S. R., Magnets and Gradients Coils: Types and Characteristics, in *The Physics of MRI: 1992 AAPM Summer School Proceedings*, American Association of Physicists in Medicine, Woodbury, New York, 1993. With permission.)

Example 7.1

Using the Biot–Savart law, estimate the magnetic field of a single-loop rectangular coil (with size a) in three-dimensional $B(r)$ space.

ANSWER

Simply, such a solution can be computed as the sum of the four line integrals based on the Biot–Savart law (Figure Ex7.1).

$$B(r) = \frac{\mu_o I}{4\pi} \oint \frac{dl \times (r - r')}{|r - r'|^3} dr \qquad \text{(Ex 7.1)}$$

$$r = xi + yj + zk \qquad \text{(Ex 7.2)}$$

$$r' = x_o i + y_o j + z_o k \qquad \text{(Ex 7.3)}$$

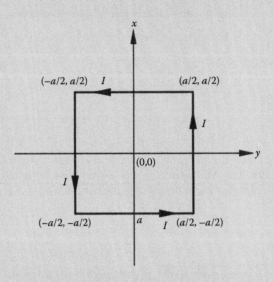

Figure Ex7.1. Application of Biot–Savart to the solution of the B_1 field for a square coil.

$$B_{top}(x_o, y_o, z) = \frac{\mu_o I}{4\pi}\left[\int_{-\alpha/2}^{\alpha/2} \frac{z_o i - \left(x - \frac{a}{2}\right)k}{\left[\left(x - \frac{a}{2}\right)^2 + \left(y - y_o\right)^2 + z_o^2\right]^{3/2}} dy \right] \qquad \text{(Ex 7.4)}$$

$$B_{bot}(x_o, y_o, z) = \frac{\mu_o I}{4\pi}\left[\int_{\alpha/2}^{-\alpha/2} \frac{-z_o i + \left(x + \frac{a}{2}\right)k}{\left[\left(x + \frac{a}{2}\right)^2 + \left(y - y_o\right)^2 + z_o^2\right]^{3/2}} dy \right] \qquad \text{(Ex 7.5)}$$

$$B_{right}(x_o, y_o, z) = \frac{\mu_o I}{4\pi}\left[\int_{\alpha/2}^{-\alpha/2} \frac{z_o j - \left(y - \frac{a}{2}\right)k}{\left[\left(x - x_o\right)^2 + \left(y - \frac{a}{2}\right)^2 + z_o^2\right]^{3/2}} dx \right] \qquad \text{(Ex 7.6)}$$

$$B_{left}(x_o, y_o, z) = \frac{\mu_o I}{4\pi}\left[\int_{-\alpha/2}^{\alpha/2} \frac{-z_o j + \left(y + \frac{a}{2}\right)k}{\left[\left(x - x_o\right)^2 + \left(y + \frac{a}{2}\right)^2 + z_o^2\right]^{3/2}} dx \right] \qquad \text{(Ex 7.7)}$$

7.4.2   Eddy Currents

Pulsation of the gradients generates a gradient fringe field that extends into the magnet cryogen compartment to induce currents, known as eddy currents, that can sometimes last for a few hundred milliseconds. In their own turn, such currents generate fields that affect the magnetic field homogeneity, and can be detrimental for image spatial resolution, and spectroscopy acquisitions.

Reduction of the eddy currents often employs active shielding, or other calibration schemes, typically carried out during quality assurance and periodic maintenance of the scanner.

7.4.3   Switching Speed

The switching speed of the gradients determines the time taken by the gradients to attain their maximum strength. The switching time is characterized by a time constant $\tau = LR$, characteristic for a small-circuit equivalent inductance (L) and resistance (R) model. Efforts to increase the field strength, by adding more windings on the gradients, increase the resistance and switching times. Effectively, efforts to increase strength and switching times are in conflict. Another term often used to characterize the performance of the gradients is slew rate ($\partial B/\partial t$).

This determines the rate of change of magnetic field strength per unit time. As discussed previously, the FDA has strict guidelines for the upper limits of the slew rates allowed for clinical imaging.

Eddy currents induced in the gradient coils are often compensated by pre-emphasis circuitry that overdrives the gradients according to the way the ideal gradient field is changed by such currents. The eddy current profile is calculated a priori in calibration testing of the scanner (Figure 7.12).

7.5   RF Coils

Radio frequency coils are used for signal reception from the magnetization in the transverse plane. RF coils are simply antennas or, alternatively, resonant LC circuits tuned at the Larmor resonant frequency. The resonant frequency is determined by

$$f_o = \frac{1}{2\pi\sqrt{LC}} \tag{7.6}$$

Another important and critical factor of their performance is the quality factor Q, often defined as the 3 dB bandwidth of their frequency response:

$$Q = \frac{f_o}{\Delta f_{\pm 3dB}} \tag{7.7}$$

RF coils are designed to receive the weak RF signal from within the human body. They are classified into transmit and receive coils. Although on certain occasions a single coil is used to transmit and receive the MRI signal (such as the body coil that is built in most scanner systems), often separate transmit and receive coils are used with connections to the transmitter and receiver RF chains, as shown in Figure 7.3. Three major types of RF coils have been introduced in the MRI field: *surface, volume,* and other specialized coils known as the *phased arrays.* Matters of importance regarding surface coils are their designs, tuning, and matching, their interconnections to the scanner, their

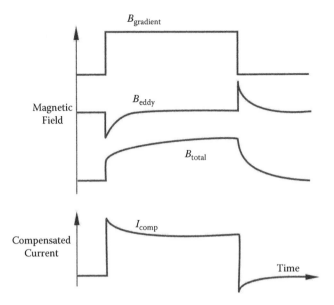

Figure 7.12. Preemphasis circuitry is often used to overdrive the gradients according to predefined patterns to offset for eddy currents and their deleterious effects. (Reproduced from Thomas, S. R., Magnets and Gradients Coils: Types and Characteristics, in *The Physics of MRI: 1992 AAPM Summer School Proceedings*, American Association of Physicists in Medicine, Woodbury, New York, 1993. With permission.)

decoupling during RF transmission and reception, and their safety. The design of surface coils is also an active area of MRI research, and numerous coil specialized designs exist.

7.5.1   Surface Coils

Surface coils are typically placed on the surface of the anatomical area of interest to receive signal proximal to their placements. Calculation of the B field distribution in free space from a conducting resonant loop indicates that there is a depth dependence of the field distribution, diminishing with distance. For human imaging, RF penetration effects need to be considered, accounting for inhomogeneous tissue structures and tissue boundaries, through solution of Maxwell's equations in the near field (where the wavelength λ is comparable to the distance of penetration). A measure of the coil's performance is its sensitivity, a function characteristic of the shape, location, and size of the coil, defined by

$$C = \oint \frac{dl \cdot r}{\left| r^2 \right|} \tag{7.8}$$

where C is the sensitivity function, dl is a displacement element along the surface of the coil, and r is the unit vector between dl and a point in space for which the field distribution is calculated.

As a rule of thumb, adequate field penetration is achieved in regions as deep as the diameter (or size) of the surface coil (Edelstein et al. 1985). Typical examples

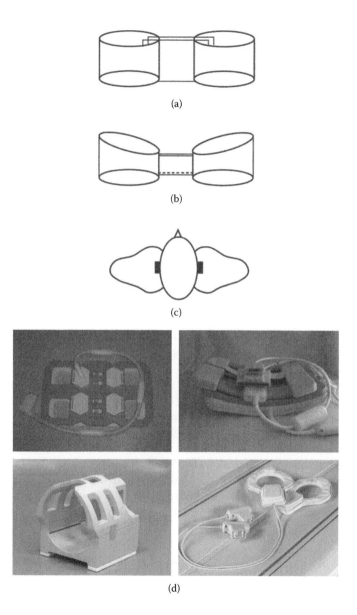

Figure 7.13. (a–c) Schematic examples of various types of radio frequency coils (breast, temporo-mandibular joint, etc.). (d) Commercially available coils.

of characteristic surface coils and an example of a prototype spiral, multiturn surface coil are shown in Figure 7.13.

7.5.2   Volume Coils: Birdcage

An extension of the type of surface coils is the volume coil. The term *volume* has been used to indicate the fact that such hardware images the anatomical volume region of interest. So, rather than placing the resonant circuit in a region proximal to the anatomy of interest, the region to be imaged is included within the coil. A typical and widespread type of volume coil is the birdcage coil (Figure 7.14; Hayes 1985). The birdcage coil is the type of coil built in most clinical scanners.

Figure 7.14. Left: A schematic of the basic design of a low-pass birdcage resonator. Middle: Its prototype counterpart. Right: Simulation of the magnetic field distribution of a birdcage coil in a mid-transverse plane in support of its superior performance in homogeneity. (Middle image is reproduced from Bronskill, M. J., and Graham, S., NMR Characteristics of Tissue, in *The Physics of MRI: 1992 AAPM Summer School Proceedings*, American Association of Physicists in Medicine, Woodbury, New York, 1993. With permission.)

It is often the preferred design for most head coils in current clinical systems. It generates a highly uniform and homogeneous field, although it has a sensitivity that is less than that of surface coils.

The birdcage coil is created from straight current-carrying elements (or segments) around a cylindrical former (Figure 7.14). The most popular designs include the low-pass and high-pass configurations, where the lump capacitor elements are placed at either the straight elements or end rings, respectively. The structure (similar to surface coils) becomes resonant due to the inherent inductance of the conducting elements. The coil can be considered an approximation to the idealized case of an infinitely long hollow cylinder. The current distribution along the z axis in each straight element follows a sinusoidal density pattern as a result of the inductor–capacitor (LC) segment delays on the end rings, given by

$$J_z = \frac{2B_1}{\mu} \sin \phi_k \tag{7.9}$$

where J_z is the current density, B_1 is the RF field, μ is the magnetic permeability, and ϕ is the angular separation between the different coil segments, given by

$$\phi_k = 2\pi k/N \tag{7.10}$$

k represents the kth straight element, and N the total number of elements in the coil. A coil with N segments generates $N + 1$ modes. The cylindrically symmetric structure of the coil generates modes in pairs. The zeroth mode is characteristic for the end ring current field and is not useful for MRI purposes. The first mode is the mode that produces the transverse field that is highly uniform. The second mode produces a field that is linearly varying. Higher-order modes produce complex fields that diminish quickly along the axis of the coil.

Birdcage coils are often driven in quadrature, with the use of a hybrid splitter circuit that drives the incoming ports with phase shifts that are 180° out of phase.

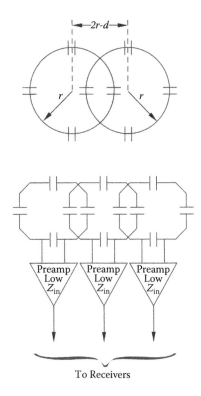

Figure 7.15. Top: The basic design of a phased array RF coil with two loops. Bottom: A typical three-loop design for a spine coil showing the blocking circuitry and receiver chain interconnection.

7.5.3 Specialized Coil Types: Phased Arrays

In 1990 Dr. Peter Roemer published an interesting scientific paper that introduced a specialized type of radio frequency antennas, known nowadays as phased arrays. The design of these coils was based on antennas used for radar communications and ultrasound. It consisted of a number of overlapping, independently receiving antennas, connected to separate low-input impedance preamplifiers (Figure 7.15) and receivers. Such arrays are receive coils and need to be decoupled from the transmit coils, a task often accomplished with the use of separate blocking circuits (Figure 7.15). Evidently, the receiver chain circuitry needs to be modified (in comparison to a single surface coil (Figures 7.3 and 7.4)), as shown in Figure 7.15. Important to the design of such arrays is the mutual coupling between the overlapping coils, which can be detrimental for resonance if it is not nulled. Surprisingly, nulling the mutual inductance can be easily achieved by overlapping the adjacent coil loops by an appropriate distance (or through the use of capacitive coupling or nonoverlapping coils with paddle end rings), as shown in Figure 7.16 for the cylindrical and planar phased arrays.

MR image reconstruction using phased arrays has traditionally been based on the sum-of-squares reconstruction, although the original work of Roemer et al. (1990) listed numerous other possible reconstruction schemes (Figure 7.17). Capitalizing and extending such concepts, recently, Sodickson and Manning (1997), Pruessmann et al. (1999), and others introduced parallel imaging reconstruction, as discussed in detail in Chapter 10.

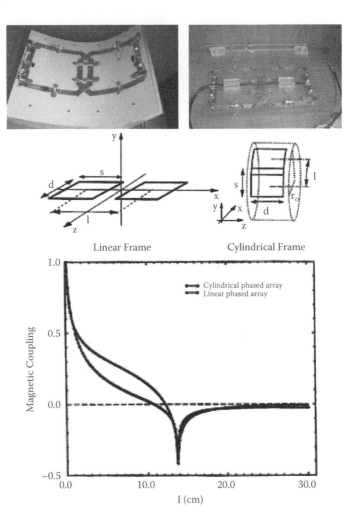

Figure 7.16. Top: Prototype cylindrical and planar two-loop phased arrays. Bottom: The mutual coupling coefficient variation with respect to the coil loop overlapping distance. (Reproduced from Constantinides, C. D. et al., *Magnetic Resonance in Medicine*, 34, 92–98, 1995. With permission.)

7.6 RF Decoupling

A critical and most important issue in the use of separate transmit and receive coils is their electrical isolation and decoupling during the two phases of the MRI experiment, that is, transmission of RF power and reception of the MRI signal. Usually, passive, high-breakdown PIN diode switches are used in parallel configurations on the RF transmit coils. Induction of voltages during the transmit phase (which can be as large as a few hundred volts) forward biases the diodes. During the reception of the weak RF signal the diodes are reverse biased and the transmitter coil effectively becomes an open circuit. More elaborate active biasing schemes are often used with a DC (direct current) voltage provided from a power supply, usually residing in the receiver cabinet. Such a scheme is used for the protection of receiver coils. The example in Figure 7.18 shows a resonant loop circuit in a phased array with blocking circuits on the upper and lower right

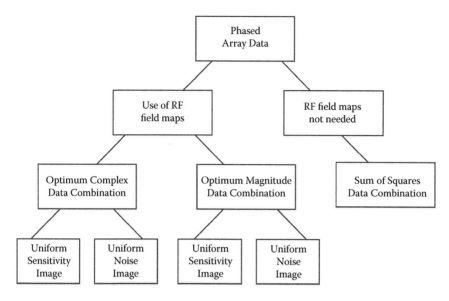

Figure 7.17. A schematic depicting the phased array reconstruction process.

ends that include a PIN diode. The diodes are biased by a DC voltage during the transmit phase. This introduces two separate resonant circuits within the resonant loop structure that causes splitting of the resonant peak at a frequency other than the transmitter frequency.

Other important connecting parts are the balun circuits often used to reduce or eliminate circulating currents, thereby providing protection to the patients (Figure 7.18).

7.7   *B* Field Distributions and Simulations

The performance of RF coils is often assessed by simulations of the magnetic field distribution in free space (and more recently, from use of dedicated high-frequency simulator software to account for the complex human and animal anatomy, electromagnetic field excitation, and tissue properties). Ideally, simulation of the field must be performed to account for the conducting medium (through the solution of Maxwell's equations), anatomy of interest, and geometry of coils. Such problems become cumbersome in view of the existence of inhomogeneous media and numerous boundaries for the penetrating RF fields, using a near-field approximation (Ramo 1984). The simulations in free space are often complemented with tests on homogeneous phantoms using the MRI scanner (Figure 7.19).

7.8   Safety Issues

There are a number of important safety issues that relate to all instrumentation used in MRI. For patient protection, the FDA has set strict guidelines (and upper limits) for a number of issues that relate to RF power deposition and specific absorption rates (SARs), patient protection during transmit and receive phases, local heating, heating of electronics, and others. Clinical scanners are equipped nowadays with protection circuitry that prevents scanning or operation of the scanner should one or more of these limits be surpassed.

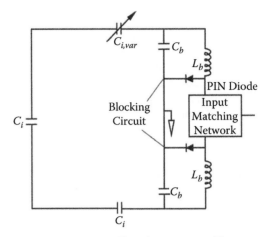

Receive Loop in Phased Array Assembly

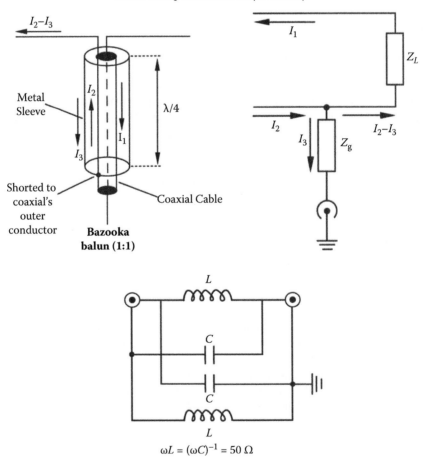

Figure 7.18. Top row: Blocking resonant circuitries (L_b, C_b) on the resonant coil with active decoupling activated upon DC bias of the PIN diodes. Middle row: A bazooka balun circuit at the connecting end of the RF antenna to minimize circulating currents. Bottom row: An alternative balun circuit using lump components.

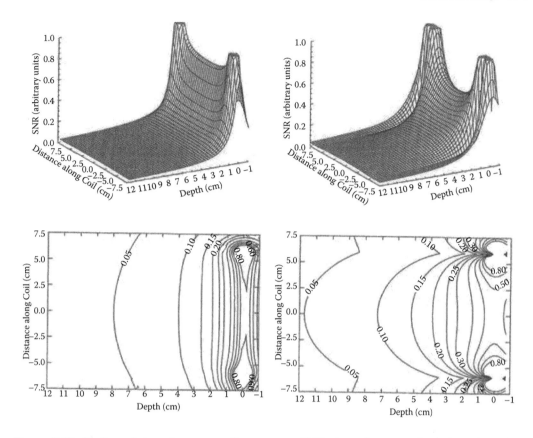

Figure 7.19. Surface plots and corresponding contour SNR plots along a sagittal plane for a (left column) planar and a (right column) cylindrical phased array (see Figure 7.17). (Reproduced from Constantinides, C. D. et al., *Magnetic Resonance in Medicine*, 34, 92–98, 1995. With permission.)

Selected Readings

1. Cho ZH, Jones J, Singh M. *Foundations in Medical Imaging.* John Wiley & Sons, 1993, New York.
2. Stark D, Bradley W. *Magnetic Resonance Imaging.* Mosby, 1999, St. Louis, MO.
3. Ramo S, Whinnery T, Van Duzer T. *Fields and Waves in Communication Electronics.* John Wiley & Sons, 1984, New York.

8 Tour of an MRI Facility

8.1 Introduction

At the very minimum a magnetic resonance imaging (MRI) facility consists of the MRI suite, the console area, a computer and hardware room, a control room, and the cryogen storage area. The filming area is usually part of the console area. In larger hospitals and clinical centers, there is often a separate room or area where an independent console or workstation physically exists that allows radiographers and radiologists to independently view, film, and process medical images.

The MRI suite is also known as the scanner room; it houses the MRI system for the clinical exams, the patient table, and a storage cabinet with all radio frequency (RF) coils and other accessories (ECG leads, phantoms, trigger or other auxiliary devices) necessary for conducting the MRI exams. In this room there are also penetration panels (metallic wall panels) that allow important electrical connections to be established with the electronics room (provides connections of preamplifiers to receivers, radio frequency amplifier to radio frequency coils, paths for DC bias voltages, etc.). The console area typically consists of the main computer that operates the scanner, and the processor for reconstruction. Often found in this area are archive devices (optical, magnetic, or digital) and media (tapes, optical or other rewritable media) for storing data. The universal image format utilized by most commercial scanners for medical images is Digital Imaging and Communications in Medicine (DICOM). A film camera, a processor, and a view box complement this area. In some cases, networking hardware exists to allow computer networking, transfer of images to workstations, and to control shared cameras.

The other support area specific to the MRI suite is the cryogen storage area. This area is designed to meet specific construction and engineering criteria to allow adequate venting, thereby preventing excessive helium and nitrogen buildup, or to allow release of vaporized gases in the case of evaporation or emergency release (quench ventilation pipes). Other support areas that complete the center include a reading room, remote viewing areas, dressing rooms, and a general waiting area for patients (Figure 8.1).

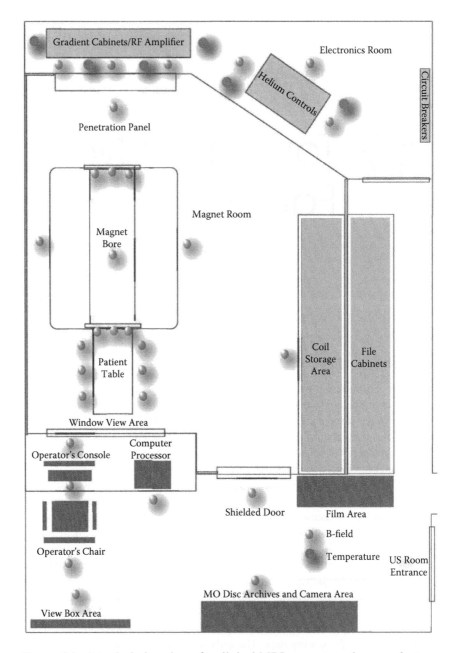

Figure 8.1. A typical plan view of a clinical MRI room, console area, electronics, and support areas. Indicative are signs for B-field and temperature mapping performed during routine quality control testing (shown as gray balls).

There are also stringent structural and environmental requirements that are considered carefully in the design and construction of an MRI suite that take into account magnet weight, vibrations, radio frequency (RF) shielding, considerations for electrical or magnetic interference from external sources, electrical grounding, monitoring and control of temperature, humidity, and air quality. There is also advanced planning for water and sewer systems. Necessary for the proper functionality of the MRI system are systems that allow oxygen monitoring and alarm systems, air quality, and adequate ventilation of the patient area within the magnet. An emergency quench mechanism is also available.

Communication systems are also carefully designed to allow intracenter communications, communications with the patient during scans, and to avoid interference of any kind with RF transmission and reception.

8.2 Hardware

The static magnetic field is oriented horizontally, and it defines the z axis. The bore has a size of approximately 0.4–0.6 m to accommodate the patient comfortably. The magnet constitutes the most expensive part of the imaging system (Figure 8.2).

The three sets of gradient magnetic fields must be positioned within the bore and are driven by three separate drivers (these are highly specialized audio amplifiers). The RF coils produce and receive the RF fields (RF transmitter and receiver). The bursts of RF energy are generated by a crystal oscillator and the pulse shaper and then amplified and delivered to the coils via the RF transmitter cabling. The elicited weak RF signals are picked up by the same or other RF coils. After amplification and demodulation of the signal by the RF receiver, the nuclear magnetic resonance (NMR) signal is sampled, digitized, and entered into the main computer.

The timing and other characteristics of the gradient fields and RF pulses are determined by the pulse programmer, a process that is under computer control. Often, when the computer is not involved with generating and acquiring resonance data, it performs the reconstruction, analysis, and processing of images for display.

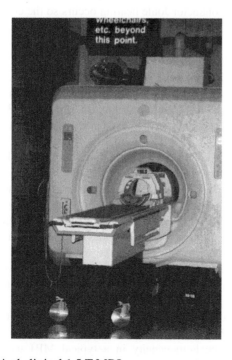

Figure 8.2. A typical clinical 1.5 T MRI scanner.

8.2.1   Instrumentation: Magnets

Typically three types of magnets are used in clinical and research MRI scanners (as described analytically in a prior section):

1. Permanent magnets

2. Electromagnets

3. Superconducting magnets

8.2.1.1   Permanent Magnets

Large blocks of ferrous metal are used to generate the magnetic field. Unlike the other types, the main field of the permanent magnet is aligned vertically. Such magnets are largely maintenance-free and consume no electric power or cryogens (i.e., liquid helium and nitrogen), but their disadvantage is that they are capable of producing only relatively low fields (up to approximately 0.4 T).

8.2.1.2   Electromagnets

Most systems are built around a superconducting magnet or an electromagnet. Both types use electrical current to generate a uniform field.

An adequate field is produced by several pairs of large-diameter coils of wire known as Helmholtz coils. The primary field is made as homogeneous as possible by moving about shim plates of steel (passive shimming) and by adjusting the small currents in various additional shim coils that produce slight field corrections (active shimming).

So for an electromagnet, copper or aluminum conductors are wound on an aluminum frame with a bore diameter of about 1 m, within which the gradient field coils and the RF coils are fitted. An inner bore of about 40–60 cm in diameter is left for the patient. Conductor Joule heating occurs so the conductors are hollow to allow circulating cooling water to cool them down.

8.2.1.3   Superconducting Magnets

Most modern machines use a superconductor. A superconducting magnet exploits the significant reduction of electrical resistance in some materials at very low temperatures (e.g., niobium-titanium alloy at −253°C). Once the current is initiated, there is no further need for power input, nor are there any conducting losses.

To maintain the superconducting condition, the entire coil must be immersed in liquid helium (approximately −269°C) contained within a cryostat (helium Dewar vessel), i.e., a stainless steel vacuum insulated cryostat that is filled with liquid helium. A surrounding refrigerator prevents warming of helium from the outside sources. Other systems use liquid nitrogen (boiling point of approximately −196°C) to prevent heat inflow to the helium container. Helium boils off slowly, so it has to be replenished every few months, and so does liquid nitrogen (even more frequently). The high cost associated with the MRI unit is mostly due to the manufacture of the magnet-cryogen assembly and its power consumption.

Other critical issues for the performance of the system are the magnet stability (which needs to be better than 0.1 ppm/h) and the field homogeneity (as defined by Equation 6.3). The homogeneity in a typical MRI unit is approximately 10 ppm over the central 50 cm bore region. The presence of any inhomogeneities within the FOV often leads to image distortion and diminution of image spatial resolution.

8.2.2 Gradient Coils

Gradient coils are also situated within the bore of the magnet, and they generate three independent sets of fields superimposed on the main static field. Such fields are spatially varying and much weaker in strength than the main static field, ranging typically from about 1 to 20 mT/m for clinical imaging. The gradient sets are usually magnetically decoupled from the main magnet through the use of a shield. Critical to the performance of the gradient sets for some imaging exams are also their slew rates (rate of change of gradient strength per unit time) and rise and fall times (times to attain a certain percentage of the maximum or minimum gradient strength).

NMR microscopy or high-field animal imaging employs fast and much more powerful gradient sets than those of a typical system to achieve smaller fields of view (FOVs) and finer spatial resolution (the voxel dimensions decrease by a factor of 100, and a resolution of about 10 μm can be achieved nowadays).

Remarkably, patients and observers are often surprised by the pulsation of the gradients and relevant noise. Such effects are the result of increased Lorentzian forces exerted on the conducting wires of the gradient coils. Ultra-fast imaging techniques, such as echo planar imaging (EPI), utilize gradient sets that can generate very high slew rates, and ultra-small rise and fall times.

8.2.3 Radio Frequency Transmission and Reception

Apart from the computer, the other major hardware components are the RF coils and electronics. The latter generate and apply the pulses at the Larmor excitation frequency and then detect the NMR signal. The implementation of this system is almost identical to a basic communication system destined for amplitude modulation (AM) broadcasting. Instead of an audio modulator, however, a pulse shaper is used to form all kinds of pulses of precise and highly specialized shapes, typically in the range of 10–64 MHz (for imaging at clinical field strengths of 1.5 and 7 T). Use of the gradients for spatial-frequency encoding imposes the use of a carrier signal that is not monochromatic, but rather contains a band of frequency components, to cover the entire spread of the spin frequencies about the Larmor frequency. It is essential that the RF generator, amplifiers, and other electronic equipment have a bandwidth sufficiently broad to cover the entire range of useful frequencies, but narrow enough to exclude most extraneous signals and, most importantly, noise. Another added complexity associated with NMR signals (as compared to AM) is that the information on the amplitude, phase, and frequency of the detected signal must be stored to be processed. Once again, RF excitation occurs via an RF transmitter that produces a weak field perpendicular to the main static field. The RF signal is then shaped (via computer control), amplified, and then transmitted to the RF coil. The excitation-detection process can take place via the same RF coil through the use of either surface or volume or surface-volume excitation. Choice of an appropriate configuration mainly depends on the imaging protocol on hand, but a common choice involves the use of the built-in volume coil for excitation and of a surface coil for reception. The reason for such a choice follows from the fact that the volume coil is best at producing a highly homogeneous excitation field throughout the FOV, and that the surface coil possesses superior signal-to-noise ratio (SNR) performance over other types of detection probes, particularly at superficial areas from its set location.

In summary, the system consists of a powerful superconducting magnet, three sets of gradients (shielded from the static field), driven by three gradient drivers, and an RF transmitter and receiver attached to the appropriate RF probe(s). All these pieces of equipment are controlled by the main computer. The entire magnet room is built within a Faraday cage (magnetic shield) to prevent interference from external signal sources and also prevent leakage of RF outside the room (Figure 8.3).

8.3   Imaging

The use of gradients allows imaging at different imaging slice orientations (axial, sagittal, and coronal as well as oblique and double oblique). This feature, in addition to the excellent soft tissue contrast it produces, renders MRI superior to computer tomography in many respects (which is confined to axial slices only). Typically, imaging at any plane is possible after positioning the patient and landmarking. This defines the isocenter point of the three-dimensional coordinate axes of the imaging setup. Prescription of a slice at any arbitrary orientation follows by defining a 3×3 rotation matrix that contains coefficients of the gradient components (G_x, G_y, G_z) to appropriately select (and subsequently excite) the prescribed plane slice in three dimensions.

8.4   Generation of MRI Images

Scanners have a built-in birdcage coil for RF excitation. Typically such a coil is also used for signal reception, unless a surface coil is chosen to be used separately. Instructive is also the placement of the shimming plates (passive) and active shimming coils for attaining increased homogeneity within the field of view and prescribed slice. Shimming is nowadays computer controlled and is often carried out during a prescan phase of the data acquisition. Water or other phantoms are also used for calibration, quality control, testing, and debugging pulse sequences or electronics, as well as for coil tuning and matching. Important also are peripheral devices that include the cardiac pacing leads and pulse oximeter to monitor patient vital signs during positioning and to provide trigger signals to the scanner for dynamic studies, such as cardiac imaging.

Once at the operator's console, the procedural steps to generate an image include selection of

1. The different imaging parameters (geometry, image dimensions, encoding, etc.)

2. Pulse sequence (acquisition scheme)

3. Reconstruction parameters

The data acquired from the RF coil are amplified via low-noise preamplifiers and boosted with the use of gain blocks before being transmitted to the receivers in the electronics room. The signal is demodulated to baseband, filtered, and then digitized. Typically, discrete Fourier transformation (DFT) is used to allow image reconstruction. Computer control (or built-in algorithms) allows visualization of the reconstructed images on the console window.

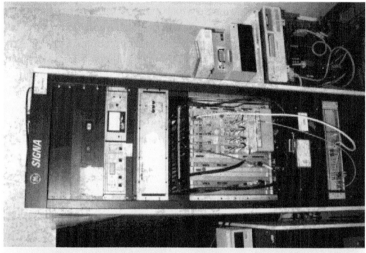

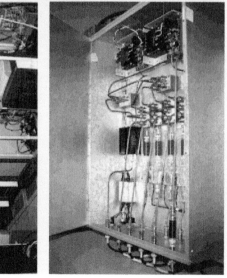

Figure 8.3. The electronics room behind the MRI scanner equipped with the three gradient cabinets: (top left) the RF exciter, (bottom left) an up/down-converter (mixer) unit, and (middle) the RF receivers.

Most commercial clinical scanners are equipped with a database for storage and retrieval of images and archiving devices (magnetic, optical, or magneto-optical drivers) and media for archiving.

8.5 Safety

There are numerous precautionary measures that international committees and other organizations (Food and Drug Administration (FDA), International Electrochemical Company (IEC)) require for safety (Shellock 2001). These cover all aspects and areas of the MRI center, which include the magnet, implanted devices, the use of metallic or other materials and auxiliary devices, RF power deposition, temperature changes during scans, use and operational limits of hardware, specific absorption rates, safety of RF coils, and others. Such safety measures are summarized in guidelines or directives published by such committees or organizations. Clinical centers are required to post signs with contraindications and warnings to ensure safety. For patient scanning, screening and consent forms from patients are also strictly required before MRI exams, as determined by institutional review boards (IRBs), ethical committees, or international bodies of ethical conduct of scientific work and practices.

9 Signal, Noise, Resolution, and Image Contrast

9.1 Signal and Noise Sources in MRI

Hoult and Richards (1976) first formulated a method of calculation of the induced signal in a nuclear magnetic resonance (NMR) experiment. Considering the radio frequency (RF) field B_1 at point P in space, produced by a coil C carrying a unit current, the presence of a rotating magnetic moment m at point P will induce an electromotive force (EMF) in the coil conductor given by

$$\text{EMF} = -\frac{\partial}{\partial t}(B_1 \cdot m) \tag{9.1}$$

$$m = M_o \, e^{-j\omega_o t} \tag{9.2}$$

Therefore, for a given sample of volume V_s, the total EMF induced is given by the equation above. For a magnetization vector rotating at the Larmor frequency ω_o, the above expression simplifies to

$$\text{EMF} = \omega_o B_1 M_o e^{-j\omega_o t} \tag{9.3}$$

where M_o is the total magnetic moment for the volume of the sample under consideration. Calculation of the signal voltage necessitates the knowledge of an expression for the magnetic field B_1 produced by a unit current flowing in the detection coil. The latter is, of course, a function of coil geometry. Therefore, an important design aspect of RF coils is the choice of such coil geometry to achieve maximization of the relative ratio of the B_1 field to the total resistance of the coil.

9.2 Noise Sources

In the course of the various topics discussed an attempt is made to characterize the physical nature of signal and noise. Noise (and signal) is often treated as a random variable, which according to probability theory characterizes random events. Therefore, both signal and noise can be characterized by a probability density function (with the mean and variance as two of the most often used statistical measures of the distributions). In magnetic resonance imaging (MRI), for example, the thermal random noise due to the motion of electrons in their detector antennas distorts the weak NMR signal that emanates from within the human body. Similarly, in ultrasound (US), propagating acoustic echoes are obstructed by hard tissue structures (e.g., bone), producing shadow artifacts of the structures that lie beneath them. Reflections of the propagating wave from reflectors other than the ones in the field of view (FOV) produce speckle noise.

The noise contributions to the detected voltage across the receiver coil in MRI come from either thermal noise in the coil or the lossy interactions between the fields generated by the coil and the sample. Below, an attempt to analyze the nature and contributions of these different noise sources is attempted.

9.2.1 Detection Coil Noise Effects

At the frequency of proton NMR imaging, phenomena such as the skin and proximity effects are associated with extra noise generation. According to the skin effect, high-frequency currents will travel primarily within a distance δ of the surface of the conductor, where δ is the skin depth.

For a long, straight cylindrical conductor, the current flow in the skin of the conducting surface yields a surface resistance:

$$r_{coil} = \left(\frac{l}{p}\right)\sqrt{\mu\mu_o\rho(T_c)\omega} \qquad (9.4)$$

where l is the conductor length, p is its circumference, μ_o is the permeability of free space, μ is the permeability of the wire, and $\rho(T_c)$ is the resistivity of the conductor.

Another effect that plays a significant role in the overall noise contribution is the proximity effect. The magnetic field created by the current of one conductor influences the distribution of current in a neighboring conductor, leading to an increase in the coil resistance, according to Equation 9.4. Other noise contributions arise due to the preamplifier (noise figure typically listed to range between 0.5 and 1.5 dB) and the Johnson thermal noise due to the electronics.

9.2.2 Sample Noise Effects

The human body is electrically and magnetically lossy. These losses have to be modeled and taken into account. The latter ultimately provide a significant limitation to the signal-to-noise ratio (SNR) performance of NMR systems. These noise voltages are induced due to the random motion of ions (e.g., Na^+, K^+, Cl^-), charged macromolecules, and fluids within the patient's body. The motion of charged macromolecules within the static field B_o produces a flickering electromagnetic field according to Faraday's law. The receiving antenna is thus almost always placed over the surface of the anatomical region of interest and often bent to conform to the geometry of this region to match the volume of sensitivity to the field of view. Irrespective of this orientation, however, the antenna will pick up a noise voltage together with the MR signal voltage. These noise voltages can be classified as dielectric and inductive.

9.2.2.1 Dielectric Losses

Such are due to the capacitive losses that occur within the sample. These are due to electric fields generated by the coil conductors (due to the flowing current in them) and the interaction of such a field on moving sample charges. The induced sample current density is governed by the equation

$$J = j\omega\left[\varepsilon_c - j\varepsilon_{lf} - j\frac{\sigma}{\omega}\right]E \tag{9.5}$$

where ω is the frequency of oscillation, ε_c is the dielectric coefficient of the material, ε_{lf} is the dielectric loss factor, σ is the conductivity, and E is the induced electric field. The dielectric coefficient can be perceived as the ability of the material to store electric fields in the form of reversible molecular motions, whereas the loss factor, ε_{lf}, expresses the tendency of the material to dissipate electric field energy through irreversible molecular motions.

The conductivity σ is a factor determining the degree to which the material can conduct currents without losses. In general, the relative magnitude of these quantities determines the type of material (dielectric or conducting) and, most importantly, the final form of the equation above.

To determine the power losses associated with a given sample volume V, the volume integral of the current density with the induced electric field can be expressed as

$$\int_V J \cdot E \, dv \tag{9.6}$$

Specifically, the distributed capacitance C_d, which is associated with every RF coil, implies the presence of dielectric and magnetic losses. To minimize such losses, high-Q capacitors and a Faraday shield must be used.

9.2.2.2 Inductive Losses

They contribute to losses associated with the conductivity of the imaged sample. This type of loss cannot be avoided, and it is thus essential to provide an estimate of its dependence and a measure of its value. Hoult and Lauterbur (1979) accounted for the inductive losses accompanying the loading of a solenoid by a human sample. They calculated that the inductive equivalent resistance of a patient with an anatomy modeled as a geometrical sphere of radius b, with conductivity σ, exposed to an RF field $B_{1,xy}$, is

$$r_{inductive} \propto \sigma\omega^2 B_{1,xy} b^5 \tag{9.7}$$

More recently, complex mathematical models have been introduced to describe the RF field penetrations in the human body, in an effort to understand further the relative effects of the permittivity and conductivity as determinant factors in the power deposition process. Although these models describe a complete mathematical analysis of planar, spherical, and cylindrical geometries to account for the difference in the shape and size of different tissues and organs in the body, the nonsymmetrical shape and anisotropic nature of organs of the human poses serious questions regarding their accuracy in human imaging.

At the proton frequency of spin precession, the sample resistance is the dominant factor in the total noise resistance. Thus, as the radius of the surface coil is decreased, the smaller sample resistance seen by the coil yields an optimum SNR. The SNR reaches a maximum, and then decreases as the size of the surface coil further decreases. This arises primarily because eventually r_{coil} will become larger than r_{sample}. This suggests that there should be an optimum coil size for a particular depth in the conducting medium (Edelstein et al. 1985). The main disadvantage of this increase in SNR by surface coils is the reduced FOV, a problem that has been recently overcome by the design of the specialized coils, known as phased arrays, exhibiting an enhanced SNR, while maintaining an increased FOV. Clearly,

$$r_{sample} = r_{inductive} + r_{dielectric} \tag{9.8}$$

The dominant role of the inductive losses, however, leads to a modified equation:

$$r_{sample} \approx r_{inductive} \tag{9.9}$$

Overall, a good indicator of the relative values of r_{coil} and r_{sample} can be determined as the ratio

$$\frac{Q_{unloaded}}{Q_{loaded}} = \left[1 + \frac{r_{sample}}{r_{coil}} \right] \tag{9.10}$$

where $Q_{unloaded}$ and Q_{loaded} are the unloaded and loaded quality factors of the RF receiver coil. Noteworthy is the fact that the coil inductance and capacitance, representing energy storage elements, generate no noise. Thermal noise is therefore due to coil and patient resistances.

9.3 Signal-to-Noise Ratio

The signal-to-noise ratio is one of the most important indices of image quality. It provides a measure of the relative intensity of the desired region of interest relative to the background noise. It is defined as the ratio of signal amplitude (or power) to the standard deviation of noise σ:

$$SNR = \frac{Signal_{ROI}}{\sigma_{bkgd}} \tag{9.11}$$

In MRI the reconstruction algorithm often becomes a determining factor toward the estimation of SNR since it leads to biases in the estimation of the background σ. In such cases, the statistical nature of noise needs to be correctly defined, after application of the nonlinear reconstruction. For example, a channel with a Gaussian distributed noise results in χ-distributed image background noise, if a sum-of-squares reconstruction algorithm is employed.

We attempt now to estimate the effects of the sample and coil losses on SNR. The induced signal in the receiver coil, according to the principle of reciprocity, is given by

$$Signal \propto V_s \omega^2 B_{1,xy} \tag{9.12}$$

where $B_{1,xy}$ is the component of the magnetic field produced by the RF coil perpendicular to B_o, V_s is the sample volume, and ω is the Larmor angular frequency of precession. The total noise detected by the coil can be attributed to the noise from the sample resistance r_{sample} and the coil resistance r_{coil} (Equations 9.4 and 9.7):

$$r_{total} = r_{sample} + r_{coil} \tag{9.13}$$

$$= [r_{inductive} + r_{dielectric}] + [r_{surface} + r_{preamp} + r_{electronics}] \tag{9.14}$$

$$\approx r_{inductive} + r_{surface} \tag{9.15}$$

$$\approx \beta\sigma\omega^2 B_{1,xy}^2 b^5 + \alpha\omega^{1/2} \tag{9.16}$$

Thus, the system SNR (per unit sample volume V_s) can be represented by

$$\psi_s = \frac{\omega^2 B_{1,xy}}{\sqrt{\alpha\omega^{1/2} + \beta\sigma\omega^2 B_{1,xy}^2 b^5}} \tag{9.17}$$

with α and β being constants of proportionality. Due to the fifth-order dependence of SNR on sample radius b, inductive losses are probably the greatest when generating images from the human torso. At low frequencies, where $\alpha\omega^{1/2} \ll \beta\sigma\omega^2 B_1^2 b^5$, the above equation reduces to

$$\psi_s \propto \frac{\omega}{b^{5/2}} \tag{9.18}$$

yielding a linear dependence of ψ_s on ω.

9.3.1 Optimizing SNR Performance in NMR Systems

The SNR performance has always been one of the most important limitations of the NMR imaging systems. In an effort to maximize the SNR for ultra-high-resolution imaging, what is first sought is the physical etiology that determines low image SNR. The fundamental limit of image SNR is set by the intrinsic SNR of a given imaging system, for a given imaging subject. This intrinsic SNR, ψ_I, is the ratio of the signal from a unit volume of sample material and the thermally generated random noise currents, induced in the sample. Effectively, ψ_I is the maximum SNR achievable for a given sample in a given NMR imaging system with a field strength B_o.

If M is the magnitude of the magnetization vector, and given that the energy difference of the spin states is much less than kT, then

$$Signal \propto \frac{dM}{dt} \propto \omega M \tag{9.19}$$

where k is the Boltzmann constant and T is the absolute temperature. Also,

$$M \propto B_o \tag{9.20}$$

$$\omega = \gamma B_o \tag{9.21}$$

A combination of Equations 9.19 to 9.21 yields

$$Signal \propto B_o^2 \qquad (9.22)$$

Equations 9.4 and 9.7 show that

$$r_{sample} \propto \omega^2 \propto B_o^2 \qquad (9.23)$$

and

$$r_{coil} \propto \omega^{1/2} \propto \alpha\sqrt{B_o} \qquad (9.24)$$

A combination of the above three equations yields

$$\psi_I = \frac{B_o^2}{\sqrt{B_o^2 + \alpha B_o^{1/2}}} \qquad (9.25)$$

Evidently, at high frequencies, ψ_I is proportional to B_o, whereas at low frequencies (when the coil resistance becomes significant compared to the patient resistance), it is proportional to $B_o^{7/4}$. The above equation clearly indicates that the highest achievable SNR is limited by the static applied field B_o. Based on the fact that a static field of 1.5 T is typically used for clinical studies nowadays, it can be realized that the maximum SNR of an MRI (excluding the options of averaging and parallel image data acquisition and processing) will always be below this upper intrinsic value ψ_I. The intrinsic SNR can be calculated from the system's SNR value ψ_s, as

$$\psi_I = \left[\frac{\psi_s \cdot 10^{NF/20}}{\left[1 - \left(\dfrac{Q_{loaded}}{Q_{unloaded}} \right) \right]^{1/2}} \right] \qquad (9.26)$$

where NF is the system's noise figure in decibels. The effort is to maximize the value of ψ_s to match as closely as possible ψ_I, by reducing the noise sources due to the preamplifier, detection system, and sample, as previously discussed.

The image SNR can be estimated by using a combination of the system SNR, the relaxation times, and the density of the tissue of interest, as well as the imaging parameters. The latter include the voxel volume, sampling time, repetition time, number of excitations, readout and phase encoding gradient amplitudes, and corresponding phase encoding steps. The image SNR can be maximized by appropriate choice of the optimum imaging parameters depending on the particular pulse sequence employed and the relaxation times of the sample tissue under consideration. For example, in an image obtained using inversion recovery or any other pulse sequences, the predicted (or voxel) SNR can be calculated by (Edelstein et al. 1986)

$$SNR_{pixel} = \psi_s \cdot A \cdot t \cdot \sqrt{T_s} \cdot \sqrt{N} \cdot f_R \qquad (9.27)$$

given a voxel of thickness t, cross-sectional area A, sampling time T_s, and number of excitations N. The variable f_R relates to the relaxation times and density of the tissue under consideration.

Example 9.1

A birdcage body coil has a quality factor (Q) of 200 when unloaded and a quality factor of 50 when loaded with a patient. If we assume that there is no inductive loading, that is, there is no change in the resonant frequency, and the measured SNR is $3000 \dfrac{\sqrt{\text{Hz}}}{\text{ml}}$, then calculate the following:

  a. The intrinsic SNR
  b. The intrinsic SNR, considering that the system's noise figure (NF) is 2.5 dB

ANSWER

a.
$$\psi_I = \frac{\psi_s}{\left(1 - \dfrac{Q_{loaded}}{Q_{unloaded}}\right)^{1/2}} = \frac{3000}{\left(1 - \dfrac{50}{200}\right)^{1/2}} = 3464 \frac{\sqrt{\text{Hz}}}{\text{ml}} \tag{Ex 9.1}$$

b.
$$\psi_I = \frac{\psi_s 10^{2.5/20}}{\left(1 - \dfrac{Q_{loaded}}{Q_{unloaded}}\right)^{1/2}} = 4619 \frac{\sqrt{\text{Hz}}}{\text{ml}} \tag{Ex 9.2}$$

9.4   Contrast-to-Noise Ratio

Equally important in image evaluation is the contrast-to-noise ratio (CNR). Such an estimate quantifies the observer's ability to distinguish and resolve neighboring tissue structures. It is defined as

$$\text{CNR} = \frac{I_{ROI_1} - I_{ROI_2}}{\left(\dfrac{I_{ROI_1} + I_{ROI_2}}{2}\right)} \tag{9.28}$$

where I_{max}, I_{min} are the maximum and minimum signal intensities within the ROI, I_{bkg} the background ROI intensity, and I_{ROI_1}, I_{ROI_2} the signal intensities of neighboring tissue structures. Decreased spatial resolution often results in poor contrast. In effect, the modulation transfer function (MTF) represents the relationship between contrast and spatial resolution, alternatively defined as the magnitude of the Fourier transfer of the system's impulse response (or equivalently, the point spread function), for positive spatial frequencies.

9.5   Tissue Parameters and Image Dependence

Image quality in MRI is dependent and characterized by a number of tissue and image parameters (McVeigh 1996). Such parameters are discussed briefly in the following two sections.

> **Proton density, $\rho(H, r)$:** This parameter represents the concentration (that is, the number of protons in a tissue volume). A proton density-weighted

image (that is, an image that is primarily based on the concentration of protons and, to a lesser extent, on other parameters) is generated when TR $\gg T_1$ (fully relaxed conditions) and TE $\ll T_2$. The repetition time is chosen to be long enough such that the magnetization has regrown to equilibrium, and the echo time is much shorter than T_2 such that little or no effective signal decay has occurred.

Spin–lattice relaxation time, T_1: The longitudinal relaxation describes the relaxation of magnetization along the z axis. After an application of an α pulse, the relaxation parameter relaxes within an interval TR according to

$$M_z(t) = M_o\left(1 - e^{-\frac{TR}{T_1}}\right) + M_z(0)e^{-\frac{TR}{T_1}} \tag{9.29}$$

T_1-weighted images are generated when TR $\approx T_1$ for that characteristic tissue, and TE $\ll T_2$, such that no significant decay in the transverse magnetization has occurred.

Spin–spin relaxation time, T_2: The transverse relaxation describes the relaxation of the magnetization in the transverse plane. Immediately after the application of an α RF pulse, the transverse magnetization is governed by

$$M_{xy}(t) = M_o e^{-\frac{TE}{T_2}}\cos\alpha + M_{xy}(0)e^{-\frac{TE}{T_2}} \tag{9.30}$$

So, for a T_2-weighted image, TR $\gg T_1$ (fully relaxed conditions) and TE $\approx T_2$ are chosen.

T_2*: In practice, T_2 is compounded by local field inhomogeneities affecting the molecular/spin scale. This leads to a faster transverse magnetization decay than that governed by T_2, such that the altered T_2, known as T_2-star, is given by

$$\frac{1}{T_2*} = \frac{1}{T_2} + \gamma\Delta B_o(r) \tag{9.31}$$

where ΔB_o represents the spatially varying local inhomogeneities. Usually, such inhomogeneities arise from the presence of metallic objects, or paramagnetic effects, leading to geometric distortions and signal falloffs.

Chemical shift, δ: This is a parameter intrinsically related to the heterogeneous structure of biomolecules, which are surrounded by respective electron clouds that "screen" the external magnetic field locally. It represents the relative frequency shift in the precessional frequency of a spin nucleus in a macromolecular structure, compared to a reference precessional frequency (e.g., that of a standard compound such as trimethylsilane (TMS)).

Another example is the magnetization of protons in fat molecules that precess slower ($\approx$220 Hz) than the magnetization of proton spins in water.

Such a phase difference manifests itself as a relative position shift of the image object in tissue compared to the normal water image. Furthermore, the observed shift is modulated by TE in gradient–echo imaging. Chemical shift is treated in more detail in the spectroscopy section (Section 10.2.1).

Velocity, v: This is moving spins in the presence of gradients during the encoding and acquisition accrue phase that, according to the magnitude and direction of the velocity, lead to signal enhancements (inflow or wash-in effects) or signal reductions (outflow or washout effects).

Magnetic susceptibility, χ_o: It defines the constant of proportionality between the external magnetic field and the induced field in tissue. There are differences in susceptibility between tissues that lead to local inhomogeneities, such as the air-tissue interface.

Diffusion coefficient, D: It characterizes the diffusion of water molecules in the presence of a magnetic field gradient. It leads to intravoxel incoherent motion that causes signal reduction. Diffusion coefficient and diffusion tensor calculation can be achieved by sensitization of spins under long-duration, high-amplitude gradients, balanced by a $180°$ pulse, in long TE acquisitions.

9.6 Imaging Parameters and Image Dependence

The imaging parameters are a set of user-defined or controlled parameters that affect signal, resolution, and contrast. Their changes are often interrelated or interdependent on the values and changes of a number of other parameters.

Repetition time, TR: This is defined as the time between successive RF pulses in a pulse sequence. In the case where TR $\gg$ TE, TR can be used to control contrast. If TR $\ll T_1$, then the signal intensity reduces with reducing TR due to diminished time for the longitudinal magnetization to relax. When TR $\ll T_2$, contrast is heavily influenced by the flip angle.

Time to echo, TE: Defined as the time between the center point of the RF excitation pulse and the center of the echo. The center of the RF pulse is determined by its shape (e.g., rectangular, sinc, hyperbolic, etc.), whereas the center of the echo in spin–warp imaging occurs at the time point when the integral under the readout gradient becomes zero. Variation in TE values offers user control of the T_2 weighting; that is, TE values of the order of the tissue T_2 are more T_2 weighted than the cases where TE $\ll T_2$. TE values also affect the sensitivity to dephasing from susceptibility-induced gradient artifacts and flow. Smaller TE values are less sensitive to such artifacts; however, shortening TE implies increased gradient amplitudes (to allow sampling of k-space in the reduced time) and a corresponding increase in the bandwidth, and thus SNR loss.

Flip (tip) angle, α: Defined as the amount of nutation or tip of the magnetization, after application of an RF pulse. In the cases of ultra-short TR values, the flip angle becomes a major determinant of T_1 contrast. The optimum flip angle (that is, the tip angle that results in the maximum image contrast or signal to noise) depends on the pulse sequence. The Ernst angle defines the optimum angle for non-steady-state gradient–echo acquisitions.

Slice thickness: The signal intensity (under conditions that resemble ideal, i.e., pulse shapes, gradient waveforms, etc.) scales approximately linear with slice thickness. Reducing slice thickness implies either decreased RF power width or increased slice selection gradient strength. In either case, the minimum TR and TE increase.

Flow compensation: These are compensation gradients (or crusher gradients) that are added on either end of bipolar gradients (e.g., on slice selection or readout gradients) to null accrued phase shifts on spins as a result of flow or motion during such gradients. Addition of such gradients increases both TR and TE.

Saturation pulses: These are played out before image acquisition and serve to suppress the signal from specific spectral moieties (e.g., spectral pulses or chemically selective pulses in fat-water suppression) or spatial pulses such as slabs placed on either side of the ventricular muscle-cavity wall to null the magnetization of inflowing spins (e.g., black blood imaging).

9.7   Resolution

The ultimate aim is the quantification of image quality and the definition of indices that will allow accurate characterization and quantification (spatial, temporal variation of indices) of medical images facilitating:

1. Diagnosis

2. Prognosis

3. Treatment planning and monitoring

4. Basic science research

Therefore, critical in the entire process is reconstruction. Reconstruction is the mathematical operation that generates/reconstructs the image. In this section resolution is discussed as one of the most important indices of image quality.

Similar to other applications, such as astronomy, spatial resolution denotes an instrument's capability to resolve two separate objects in space, time, or energy as separate. It is the minimum distance or time between adjacent objects or events in order for the radiologist or the reviewer to perceive them as separate.

A simple way to understand spatial resolution is to consider two point sources of finite width and to hypothesize that a medical imaging system behaves as a linear shift invariant (LSI) system (Figure 9.1).

Imaging a point object (having an infinitesimal width) with a medical imaging system results in an image with finite extent/width—known as the impulse response $h(x, y)$ or line response $l(x, y)$ function (Figure 9.2).

We use the term full width at half maximum (FWHM) to quantify the resolution of a medical imaging system. In effect, FWHM is the width of one of these profiles at one-half its maximum value.

Typically, in medical instruments or devices, the resolution is anisotropic; i.e., it differs from one dimension to the other. A characteristic example is the case of ultrasound imaging where resolution is by far enhanced along the transducer axis

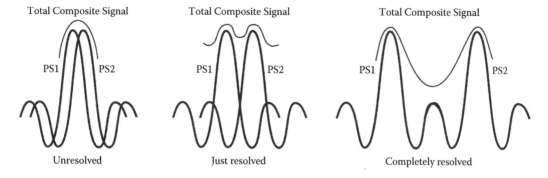

Figure 9.1. Illustration of the principle of resolution using two point signal sources (PS1, PS2).

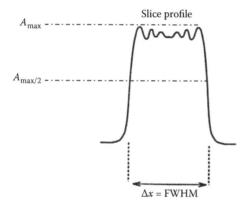

Figure 9.2. Definition of effective resolution of a medical imaging system from a slice profile of a point object image.

compared to its resolution in the orthogonal direction. In effect, the total resolution in an LSI system is the convolution of the impulse responses of each of the contributing parts, i.e.,

$$g(x, y) = h_1(x, y) ** h_2(x, y) ** h_3(x, y) ** f(x, y) \qquad (9.32)$$

$$= h_T(x, y) ** f(x, y) \qquad (9.33)$$

Effectively, the total resolution R_T is given by

$$R_T = \sqrt{R_I^2 + R_E^2 + R_P^2} \qquad (9.34)$$

where R_I is the intrinsic detector resolution, R_E is the resolution due to the system's extrinsic factors (acquisition and data collection schemes and reconstruction), and R_P is the resolution due to physiologic motion or other patient factors (peristaltic motion, breathing, cardiac motion, etc.).

Example 9.2

Prove that the total resolution of the LSI system described above is given by Equation 9.34.

ANSWER

Assume that the total system response, G, is governed by the convolution of the responses due to intrinsic (h_I), extrinsic (h_E), and physiological factors (h_P), collectively represented by the total impulse response h_T, as follows:

$$G = h_I **h_E **h_P **I = h_T **I \qquad \text{(Ex 9.3)}$$

$$h_T = h_I **h_E **h_P \qquad \text{(Ex 9.4)}$$

$$h_i = A_{o,i}e^{-t^2/\sigma_i^2} \overset{FT}{\Leftrightarrow} A_{o,i}e^{-\pi^2 f^2 \sigma_i^2} \qquad \text{(Ex 9.5)}$$

$$G(\omega_x,\omega_y) = H_I(\omega_x,\omega_y) \cdot H_E(\omega_x,\omega_y) \cdot H_P(\omega_x,\omega_y) \qquad \text{(Ex 9.6)}$$

$$= A_{o,I}e^{-\pi^2 f^2 \sigma_I^2} \cdot A_{o,E}e^{-\pi^2 f^2 \sigma_E^2} \cdot A_{o,P}e^{-\pi^2 f^2 \sigma_P^2}$$

$$= \left(A_{o,I}A_{o,E}A_{o,P}\right)e^{-\pi^2 f^2 \left(\sigma_I^2 + \sigma_E^2 + \sigma_P^2\right)} \qquad \text{(Ex 9.7)}$$

$$\sigma_T^2 = \sigma_I^2 + \sigma_E^2 + \sigma_P^2 \qquad \text{(Ex 9.8)}$$

$$e^{-\pi^2 f_{1/2}^2 \sigma_T^2} = \frac{1}{2} \qquad \text{(Ex 9.9)}$$

$$2f_{1/2} = \text{FWHM} = 2\sqrt{\frac{\ln 2}{\sigma_T^2}} \qquad \text{(Ex 9.10)}$$

Selected Readings

1. Cho ZH, Jones J, Singh M. *Foundations in Medical Imaging*. John Wiley & Sons, 1993, New York.
2. Edelstein WA, Glover GH, Hardy CJ, Redington RW. The Intrinsic Signal-to-Noise Ratio in NMR Imaging. *Magnetic Resonance in Medicine* 1986; 3(4):604–618.
3. Purcell E, Torrey HC, Pound RV. Resonance Absorption by Nuclear Magnetic Moments in a Solid. *Physical Review* 1946; 69:37–38.

⑩ Spectroscopy and Spectroscopic Imaging

10.1   Introduction to NMR Spectroscopy

Nuclear magnetic resonance (NMR) spectroscopy is the study of biomolecules by recording the interaction of an applied radio frequency (RF) field with the nuclei of such molecules, in the presence of a strong external magnetic field. Zeeman first observed the behavior of certain nuclei subjected to a strong magnetic field at the end of the last century, but practical use of the so-called Zeeman effect (refer to Chapter 2) was only made in the 1950s when NMR spectrometers became widely available.

After the first successful NMR experiments, performed independently by Purcell, Torrey, and Pound (1946), and Bloch, Hansen, and Packard (1946), it was shown that the NMR in bulk materials can be observed in several ways, including the slow passage experiment that involves a sweep of an applied RF field in the presence of an external magnetic field and the subsequent collection of the resonant signal as a series of lines with a shape close to that of the Lorentzian function. This technique, together with an adiabatic rapid passage, is referred to as a continuous wave (CW) technique, since the RF is applied in a continuous manner, while the spectrum is observed. A third method, proposed initially by Hahn (1950), uses "short bursts" of RF power at discrete frequencies to excite the population of nuclear spins. Upon completion of the irradiation, the spectrum is observed.

Although both the CW and pulsed methods were introduced and improved since the middle of this century, pulsed methods are the ones widely employed nowadays. Such methods were in fact really appreciated by the research community as soon as the power of the Fourier transform (FT) and its applicability in NMR was first realized by Ernst and Anderson (1966).

The tremendous growth of structural chemistry and biochemistry in association with the advent of digital electronics triggered the realization of new, more sensitive spectroscopic techniques implemented with faster, inexpensive, and more sophisticated computers and superconducting magnets. New ingenious experiments, tools, and techniques have been constantly emerging, establishing NMR as one of the most powerful research techniques for the elucidation of the mysteriously complex biological mechanisms and reactions of the molecular microcosm.

Interest in high-resolution MR studies in biological tissues started in 1973 with the study of red blood cells (Moon 1973), and gained fruitful grounds in 1974 when Hoult et al. (1974) showed that ^{31}P spectra could be obtained from the calf muscle.

The first part of this chapter presents a succinct introduction to the basic physical concepts that underlie NMR spectroscopy. It discusses the use of localized spectroscopy and the numerous problems one often encounters in its everyday use. Remedies are proposed wherever possible, drawing on knowledge from the plethora of existing literature publications (Bottomley 1989a).

The second part introduces the reader to the use of advanced NMR techniques for spectroscopy and spectroscopic imaging.

Although NMR spectroscopy is deeply rooted in the theory of quantum mechanics—from which it initially emerged—the classical approach is used to describe the concepts in this treatment. The interested reader is referred to the classical textbooks by Abraham (1972) and Slichter (1990) for a detailed treatment of NMR based on quantum mechanics.

10.2   Fundamental Principles

NMR is a unique tool in quantifying metabolic concentrations and other information of biochemical parameters. Phosphorus (^{31}P), for example, provides information such as the cellular energy state and intracellular pH, as well as details on phospholipid metabolism. Water-suppressed proton (^{1}H), on the other hand, allows quantification of various intermediate metabolites, including amino acids and lactate.

10.2.1   Chemical Shift

Spectroscopy is inherently linked to chemical shift, the intrinsic property of biomolecules and chemical moieties to precess at different frequencies within the chemical environments in which they exist. Chemical shift is the relative difference in the precessional frequency of different parts of a given biomolecule (or different molecules) with respect to a reference precessional frequency (Proctor and Yu 1950). The explanation of these small shifts is found in the atomic electrons. Around every nucleus there are a number of circulating electrons, often part of a bonding cloud that binds the atoms together into molecules. Such electrons generate a weak magnetic field with a diamagnetic nature (opposing the strong external magnetic field), thereby screening the nuclei from the external magnetic field B_o. The proton nuclei experience in this way a slightly smaller magnetic field than the net external field B_o. The effective net magnetic field is

$$B_{eff} = B_0(1-\sigma) \tag{10.1}$$

leading to an effective resonant frequency of

$$f_o = \frac{\gamma B_{eff}}{2\pi} = f_\delta(1-\sigma) \tag{10.2}$$

where σ is a shielding tensor (anisotropic in bound molecules and isotropic in solution). This shielding parameter is defined in parts per million of the resonance frequency:

$$\sigma = \frac{(f_o - f)}{f_o} \cdot 10^6 \qquad (10.3)$$

It is often easier to refer to chemical shifts relative to standard compounds such as trimethylsilane (TMS), phosphoric acid (H_3PO_4), phosphocreatine (PCr), and others. TMS is in general used as a reference compound:

$$\delta = \frac{(f_{TMS} - f)}{f_{TMS}} \cdot 10^6 \qquad (10.4)$$

Chemical shift is a fundamental property since it allows the same nuclei or compounds to be distinguished in different chemical or molecular environments. For example, a classical example of chemical shift is the difference in resonant frequencies of water and fat protons (Figure 10.1). This is due to the fact that the methyl (CH_3-) protons in fat are screened by a different electron density distribution than that of water. Another typical example of a spectrum is that of phosphorus (^{31}P) from muscle. Spectra of this kind provide a noninvasive method for assessing the metabolic state of the tissue in both health and disease (Figure 10.2).

While σ is a constant, the chemical shift increases linearly with field strength, so the spectral resolution (and signal-to-noise ratio (SNR)) increases at higher field strengths.

10.2.2 Theory

Lamb (1941) calculated in closed form the shielding factor σ for isolated atoms:

$$\sigma_D = \frac{e^2}{3mc^2} \int \frac{\rho(r)}{r} dr \qquad (10.5)$$

where r is the radial distance from the nucleus, c is the speed of light, m is the nuclear mass, e is the electron charge, and $\rho(r)$ is the charge density. This expression was, however, inadequate to explain the restrictions in motion of electrons

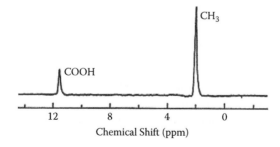

Figure 10.1. The ^1H spectrum of acetic acid (CH_3COOH). The relative signal intensities of the two peaks have a ratio of 1:3. The frequencies of the two moieties are expressed in ppm (relative to the signal from TMS). (Reproduced from Gadian, D.G., *Nuclear Magnetic Resonance and Its Applications to Living Systems*, Clarendon Press, Oxford, 1982. With permission.)

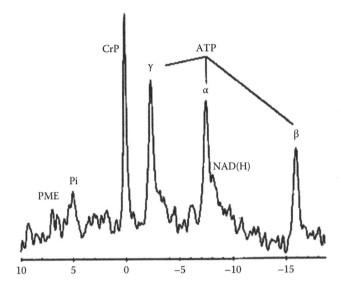

Figure 10.2. Typical ^{31}P NMR spectrum from an isolated rat heart. PME, phosphomonoesters; Pi, inorganic phosphate; CrP, phosphocreatine; ATP, adenosine triphosphate. (Reproduced from Deslauriers, R. et al., Spectroscopy: Principles and Additional Instrumentation, in *The Physics of MRI: 1992 AAPM Summer School Proceedings*, American Association of Physicists in Medicine, Woodbury, New York, 1993. With permission.)

in a biomolecule. Following Lamb, Ramsey (1950) introduced a paramagnetic term σ_P (negligible for ^{1}H). Additionally, other shielding mechanisms exist that contribute to the shielding factor, leading to a composite value:

$$\sigma = (\sigma_D + \sigma_P + \sigma_A + \sigma_R) \tag{10.6}$$

where σ_A accounts for interactions with neighboring nuclei and σ_R accounts for aromatic ring or Π-electron systems.

In molecules, chemical shift is an anisotropic tensor quantity with the shielding factor in liquids being represented by an average isotropic shielding constant σ,

$$\sigma = \frac{1}{3}(\sigma_{11} + \sigma_{22} + \sigma_{33}) \tag{10.7}$$

Two additional important aspects of NMR spectroscopy are sensitivity and resolution. The intensities of NMR peaks, as measured from their areas, are proportional to the number of nuclei that contribute toward them. Other factors, such as the spin–lattice relaxation time T_1 and the nuclear Overhauser effect, can also affect signal intensities. The line width of spectra is given by the full width at half maximum (FWHM) distance (Gadian 1976), evaluated according to the methodology presented in Section 9.7 (Figure 9.2).

The width of spectra thus depends on T_2, T_2*, but also on mobility of species (highly mobile, sharper peaks) and chemical exchange (Figure 10.3; Gadian 1976).

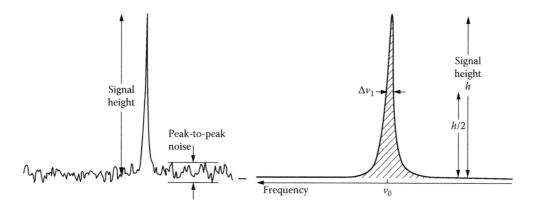

Figure 10.3. Left: Measuring signal to noise in an NMR spectrum. The Lorentzian line shape. Resolution is measured as the full width at half maximum. (Reproduced from Gadian, D. G., *Nuclear Magnetic Resonance and Its Applications to Living Systems*, Clarendon Press, Oxford, 1982. With permission.)

10.2.3 Resolution in Spectroscopy

It would be instructive to consider the resolution (often defined as the FWHM) of spectra as well as the effect of the magnitude operation on these. For the absorption, $A(\omega)$, or dispersion, $D(\omega)$, line widths,

$$Max\|A(\omega)\| = T_2 \tag{10.8}$$

So the half maximum point occurs at

$$\frac{\dfrac{1}{T_2}}{\left(\dfrac{1}{T_2}\right)^2 + \omega_{HM}^2} = \frac{T_2}{2} \tag{10.9}$$

Rearrangement of the above equation yields

$$\omega_{HM} = \pm \frac{1}{T_2} \tag{10.10}$$

or equivalently

$$\Delta f = \text{FWHM} = \frac{1}{\pi T_2} \tag{10.11}$$

which is simply the spectral resolution. In the case of a magnitude spectrum, the power spectrum $P(\omega)$ is defined as

$$P(\omega) = \sqrt{[A(\omega)]^2 + [D(\omega)]^2} \tag{10.12}$$

$$= \frac{1}{\sqrt{\omega^2 + (1/T_2)^2}} \tag{10.13}$$

In a similar manner, it can be shown that

$$\Delta f = \text{FWHM} = \frac{\sqrt{3}}{\pi T_2} \qquad (10.14)$$

The equations above indicate that resolution diminishes with the magnitude operation.

10.2.4 Spin–Spin Coupling

This phenomenon gives rise to the multiplet structure of spectra. Such a coupling arises from interactions between nuclei that cause energy level splittings, giving rise to several transitions instead of the single energy transition. Spin–spin coupling is an intramolecular effect (a through-bond effect). The coupling constant J (in Hz) denotes the frequency difference of the different multiplet peaks, and it is independent of the external magnetic field B_o. A first-order analysis can be used to predict the spectral multiplet pattern if $\Delta f \gg J$ and the nuclei (or groups of nuclei involved) are both chemically and magnetically equivalent.

The second kind of splittings due to internuclear interactions is the dipole-dipole interactions that depend on the angle of the vector that joins two dipoles, and B_o. These interactions are intermolecular; they fall off with $1/r^3$, and average out to zero for isotropic motion of molecules in fluids.

10.2.4.1 Mechanism of Spin–Spin Coupling

The most energetically favored state between two neighboring spins A and B is when both are antiparallel. When the magnetization moment of A is inverted with respect to B, during an NMR experiment, two spectral peaks result. The difference in their frequency depends on J, the interaction or coupling between the two spins. Coupling can exist between neighboring spins that are bonded together or between spins that are several bonds apart. The degree of coupling can be assessed from the comparison of the difference in chemical shifts of the two nuclei, with respect to J. When the difference in the chemical shifts $|\delta_A - \delta_B| \gg J$, this is referred to as the weak coupling case. When this rule is no longer valid, the spectra are said to be strongly coupled, or second order coupled. Coupling can be homonuclear or heteronuclear, i.e., between the same nuclei or between different nuclei (Gutowsky 1951). The interaction between neighboring nuclei is reciprocal; that is, nucleus A splits nucleus B, and vice versa.

For weakly coupled spins, the pattern of spectral peaks is characteristic of a *multiplet*. The number of peaks, spaced in frequency and relative intensities, result in the following few rules (Figure 10.4):

- A nucleus (or a set of nuclei) coupled to a set of n other nuclei will exhibit a multiplet pattern with $2nI + 1$ spectral peaks, where I is the spin quantum number

- For spins with $I = 1/2$ the intensities of the $n + 1$ spectral peaks in the multiplet are given (in relative terms) by the coefficients of a binomial expansion according to Pascal's triangle

- The $2nI + 1$ multiplets are equally spaced and separated by the coupling constant J

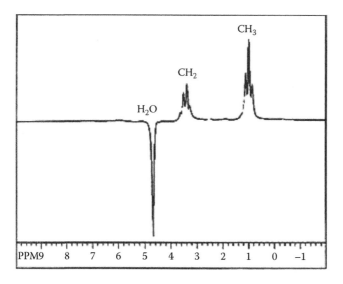

Figure 10.4. Spectrum of ethanol with multiplet peaks. (Reproduced from Barker, P., MRI Lecture Notes, Johns Hopkins University, Baltimore, 1992. With permission.)

In the case of strongly coupled spins, multiplet symmetry is lost and no longer follows the basic rules listed above. The splittings are no longer equal to the coupling constant. New spectral lines (known as combination lines) appear, and spectral peaks do not uniquely identify a particular nucleus anymore. The spectrum in these cases can only be determined by simulation. For nuclei with $I > 1/2$, the spin couplings become complicated. Such treatment is beyond the scope of this book.

10.2.5 Decoupling and Nuclear Overhauser Effect (NOE)

Irradiation of a sample by an RF field of sufficient strength can lead to a collapse of the multiplet pattern into a single line if the coupling nucleus changes its alignment relative to B_o, at a rate faster than J_{AB}. This is what is referred to as spin decoupling and leads to an increase in the SNR of the spectrum. The multiplet pattern collapses to a singlet with a total spectral area equal to the area of the individual spectral peaks in the multiplet. In the case where the nuclei are the same, this is referred to as homonuclear decoupling. In the case where the nuclei are different, the phenomenon is referred to as heteronuclear decoupling.

The Nuclear Overhauser Effect (NOE) is another useful method for increasing the SNR of nuclei other than protons, when protons are coupled to them. Irradiation of the protons attached to nucleus B, using an RF field, causes a change in the population of the energy levels of the nucleus under observation, A, via dipole–dipole coupling between A and B. The signal enhancement is

$$\frac{S_{B_{sat}}}{S_{no,sat}} = 1 + \frac{\gamma_B}{2\gamma_A} \tag{10.15}$$

where γ_A and γ_B are the gyromagnetic ratios of nuclei A and B, respectively. The NOE is the manifestation of an intermolecular effect that is a through-space effect

(Overhauser 1953). Interestingly, this phenomenon was discovered by Overhauser, a young student, during his postgraduate work.

10.2.6 Solvent Suppression

Solvent suppression techniques are used on a number of occasions to suppress particular undesired metabolites. Some useful solvent suppression techniques include the following:

1. Inversion recovery and sampling at time $t = TI$, known as the inversion time

2. Use of long echo time TE to attenuate signal from short T_2 moieties

3. Chemical saturation (using spectrally selective pulses) of a particular moiety or chemical excitation of a particular spectral moiety

4. Use of spectral (or quantum) filtering techniques

10.3 Localized Spectroscopy

In spectroscopy the inherent frequency distribution of the NMR spectrum needs to be maintained. If an object that consists of multiple metabolite moieties that exhibit different chemical shifts is imaged using frequency encoding gradients, the resulting composite image will be the spatial superposition of the individual images of each moiety, shifted in the frequency encoding direction by an amount determined by the chemical shift of the constituent moieties and the value of the encoding gradient. In ^{1}H MRI, where two dominant chemical moieties exist (water and fat) that resonate with a frequency difference of only 220 Hz apart (at 1.5 T), the composite image consists of the image of the water distribution superimposed on a shifted replica of the same image along the frequency encoding direction (spin–warp imaging). The resulting artifact is known as the chemical shift artifact. Therefore, the only difference of spectroscopy with imaging is that frequency encoding cannot be used for localization.

Another issue is the slice selection. To ensure that the same moieties within the slice are excited regardless of their chemical shift differences, the slice selection gradient must be much larger than the maximum chemical shift difference among them. Interestingly, the evolution of spectroscopy progressed initially with the advent of a number of spectroscopic techniques that addressed this issue. Recently, there has been resurgence in the development of spectroscopic imaging techniques and direct imaging techniques of nuclei other than protons. In the development of such techniques, great efforts have been expended to address and deal with the problems of imaging multiple nuclei exhibiting differences in chemical shifts.

Most existing spectroscopic techniques depend upon their ability to localize data acquisition to the region of interest (critical in focal disease) and on the quality of the achievable spectral resolution. Adequate resolution can be achieved by optimizing the magnetic field gradient homogeneity over the sample volume under investigation. One way to achieve this is by limiting the volume of interest by some form of volume selection/localization mechanism. A number of different volume selection methods have been developed, including rotating-frame zeugmatography (RFZ), depth-resolved surface coil spectroscopy (DRESS), point-resolved surface coil spectroscopy (PRESS), 1D, 2D, and 3D chemical shift

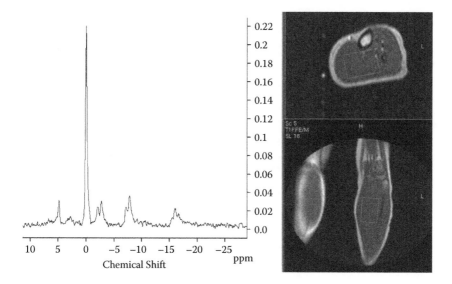

Figure 10.5. ^{31}P spectrum of the calf muscles (left) using a graphic voxel local-ization scheme based on a high-resolution ^{1}H anatomic scout image (right). (Reproduced from GyroTools, Inc., Lecture Notes, ETH, Zurich. With permission.)

imaging (CSI), stimulated echo acquisition mode (STEAM), and image-selected in vivo spectroscopy (ISIS). In the following section, these and other localization methods are explained (Figure 10.5).

10.3.1 Surface Coils

One of the simplest methods utilized early on to achieve spatial localization was surface coils. Numerous spectroscopy techniques took advantage of the sensitivity profile of the coil at planes perpendicular to the cylindrical axis of the coil, as well as the fact that its B_1 field falls off with depth inside the tissue (refer to Chapter 7). The sensitivity falloff pattern provides the basis for local selectivity/localization in a direction perpendicular to the coil's axis.

10.3.2 Depth Localization and Localized Spectroscopy

The following are some spectroscopic techniques for spectroscopic acquisitions. The advantage of most of these techniques is that they are pulse-acquired tech-niques, and thus not T_2 sensitive.

10.3.2.1 DRESS

DRESS uses slice selection to excite spectra in a thin section parallel to the surface coil. Signal is acquired in the absence of a gradient. SLIT DRESS is an extension of DRESS that allows multiple slice acquisition, using interleaved selective slice excitations (Figure 10.6).

10.3.2.2 Rotating Frame Zeugmatography

This technique encoded spectra at various depths into the tissue at multiple sec-tions parallel to the plane of the surface coil, taking advantage in this way the nonuniform pattern of the field distribution of the surface coil.

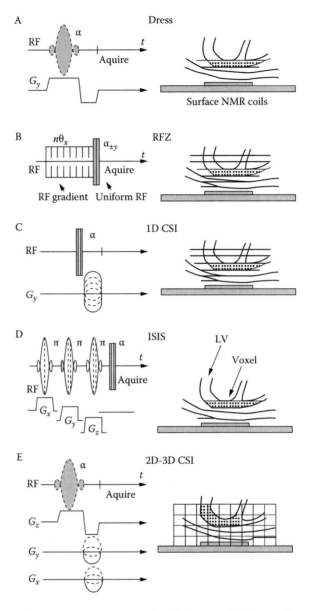

Figure 10.6. Numerous spectroscopic data acquisition pulse sequences. (Reproduced from Bottomley, P. A., *Radiology,* 191(3), 593–612, 1994. With permission.)

10.3.2.3 1D-CSI

One-dimensional (1D)-CSI uses a nonselective RF pulse in association with a phase encoding gradient oriented perpendicular to the surface coil to spatially encode multiple sections parallel to the coil.

10.3.2.4 2D- or 3D-CSI

This technique uses slice selection to select an axial slice and then phase encodes information in the remaining two or three dimensions. Three-dimensional (3D)- CSI requires a large number of phase encoding steps for the entire sample. This imposes a time restriction/limit to its use for dynamic studies such as cardiac studies.

10.3.2.5   PRESS

In PRESS the second and third 180° pulses are slice selective (Figure 10.7). Crusher pulses are used to completely dephase any transverse magnetization (not shown in Figure 10.7).

10.3.2.6   STEAM

The STEAM sequence (Figure 10.8) uses a second 90° pulse to invert and allow regrowth of the longitudinal magnetization within a time interval (between the second and third RF pulses), known as the mixing time (TM), thereby preserving or modulating T_1 contrast. The first and third 90° pulses are equally spaced in time and last for TE/2 ms. The evolution of the magnetization vector (and spin ensemble thereof) is complicated and beyond the scope of this chapter. The interested reader is referred to the book by Cho (1993).

10.3.2.7   PRESS and STEAM

PRESS has an increased SNR performance but poorer spatial localization (since the 180° pulses do not form as sharp of a slice profile as the 90° pulses) than STEAM. STEAM also allows observation of short T_2 nuclei. Since both are based on echo refocusing, they are T_2 sensitive.

10.3.2.8   ISIS

ISIS uses eight slice-selective MRI inversion pulses to localize to a single voxel. Each selective excitation (in the three orthogonal planes) is followed by

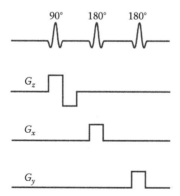

Figure 10.7. Point-resolved surface coil spectroscopy (PRESS) pulse sequence.

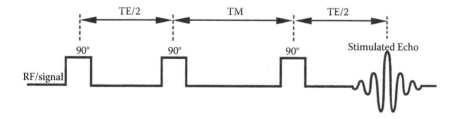

Figure 10.8. Stimulated echo acquisition mode (STEAM) pulse sequence.

a nonselective excitation, and by data acquisition. Because eight acquisitions are required for isolating the spectrum, this technique becomes susceptible to motion and other artifacts.

10.4   Imaging Equation and Spectroscopic Imaging

In Fourier transform (FT) NMR, the sample is irradiated with a short, intense, polychromatic burst of RF power. The result is that all nuclei in the sample will emit at their characteristic NMR frequencies simultaneously. In the simplest case, where only one nucleus is in the sample, a decaying cosinusoidal signal is obtained (which decays with T_2). In typical biological solutions where multiple nuclei exist, each having a different intensity and resonating at different frequencies, the free induction decay (FID) obtained from such an experiment is given by

$$f(t) = \sum_{j=1}^{k} e^{i\omega_j t} e^{-\lambda_j t} \tag{10.16}$$

where $\lambda_j = \lambda = 1/T_2$ in a homogeneous field and excited spins with the same T_2 response, and ω_j is the jth resonant frequency in the spectrum. The time domain signal (FID) can also be decomposed into its spectral components by invoking the Fourier series theorem:

$$f(t) = \sum_{n=1}^{N} A_n \cos\left(\frac{n\pi}{T}\right)t + \sum_{n=1}^{N} B_n \sin\left(\frac{n\pi}{T}\right)t \tag{10.17}$$

$$= \sum_{n=1}^{N} C_n e^{i\frac{n\pi t}{T}} \tag{10.18}$$

where $C_n = \sqrt{A_n^2 + B_n^2}$ and $\omega_n = \dfrac{n\pi}{T}$. The spectrum is simply obtained by taking the Fourier transform of the time domain signal:

$$F(\omega) = FT[f(t)] \tag{10.19}$$

$$= \int_{-\infty}^{+\infty} \left[\sum_{j=1}^{k} e^{-\frac{t}{T_2}} e^{i\omega_j t} \right] e^{-i\omega t} \, dt \tag{10.20}$$

$$= \sum_{j=1}^{k} \int_{-\infty}^{+\infty} e^{-t[\lambda + i(\omega - \omega_j)]} dt \tag{10.21}$$

$$= \sum_{j=1}^{k} \left[\frac{1}{[\lambda - i(\omega_j - \omega)]} \right] \tag{10.22}$$

$$= \sum_{j=1}^{k} \left[\frac{\lambda + i(\omega_j - \omega)}{\lambda^2 + (\omega_j - \omega)^2} \right] \tag{10.23}$$

$$= \sum_{j=1}^{k} \left[\frac{\lambda}{\lambda^2 + (\omega_j - \omega)^2} + i \sum_{j=1}^{k} \frac{(\omega_j - \omega)}{\lambda^2 + (\omega_j - \omega)^2} \right] \tag{10.24}$$

$$= A(\omega) + iD(\omega) \tag{10.25}$$

where $A(\omega)$ and $D(\omega)$ are known as the absorption and dispersion line shapes. In real spectrometers, the time domain FID is first sampled before it is Fourier transformed. Starting from Equation 10.25, derived above, the amplitude of the jth point of the frequency spectrum of an FID can be given by

$$F(\omega_j) = \sum_{l=0}^{N-1} C_n(t) \cdot e^{-\frac{\pi j l}{N}} \tag{10.26}$$

where $F(\omega_j)$ is the amplitude of the jth point in the frequency domain, $C_n(t)$ is the amplitude of the lth point in the time domain, and N is the number of data points.

10.4.1 Frequency-Selective Pulses: Frequency Selection

Frequency selection is achieved by application of an RF pulse in the presence of a slice selection gradient G_z. If G_z is assumed to be fixed, then the RF pulse imparts small tip angles (refer to Pauly's selective excitation paper (Pauly et al. 2001)), similar to imaging:

$$\omega(r) = \gamma (B_o + G \cdot r) \tag{10.27}$$

$$= \gamma \left(B_o + \frac{\partial B_z}{\partial r} \cdot r \right) \tag{10.28}$$

$$= \gamma \left(B_o + \frac{\partial B_z}{\partial z} \cdot z \right) \tag{10.29}$$

In the rotating frame, given the existence of the moiety's chemical shift σ:

$$\omega = \omega(r) - \gamma B_o \tag{10.30}$$

$$= \omega(r) - \omega_o \tag{10.31}$$

$$= \gamma [\sigma + G(r).r] \tag{10.32}$$

Thus, the frequency depends on both the chemical shift and the position of the spins. Hence, the position can be computed from

$$r = \frac{\left(\dfrac{\omega}{\gamma} - \sigma \right)}{G(r)} \tag{10.33}$$

Evidently, the relationship between position and chemical shift is linear. The slice thickness can thus be expressed as

$$ST = \frac{BW_z}{\gamma G_z} = \frac{1}{\gamma G_z \Delta t_{rf}} \qquad (10.34)$$

To ensure that all precessional frequencies of the selected moieties are excited, the bandwidth (BW) has to be adequately high, and so must be the gradient strength. For nuclei other than ^1H, problems are encountered since the chemical shift dispersion is very high. Problems may also be encountered in cases where multiple thin slices are excited.

10.4.2   Quantification

Absolute quantification of metabolites requires nontrivial acquisition protocols that often involve phantoms and instrument calibrations and relaxation time measurements. They are based on the use or external or internal reference markers (within the field of view of the scout image) on which the unknown quantity is referred to (Figures 10.9 to 10.11).

10.4.3   Spectroscopic Imaging

The term *spectroscopic imaging* emerged as the generic name to the generation of images from spectra after spectral editing and filtering. A number of computer programs were developed to allow spectra manipulations and processing (see Figures 10.12 and 10.13). Recently, there has been resurgence in the interest for direct imaging and quantification of biologically important and relevant nuclei (such as ^{23}Na, ^{39}K, ^{31}P, ^{13}C, and others).

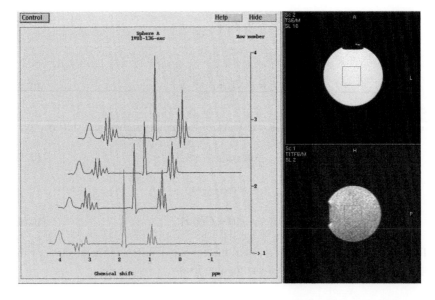

Figure 10.9. ^1H spectra acquisition from a uniform cylindrical phantom doped with different chemical moieties, using spatial localization. (Reproduced from GyroTools, Inc., Lecture Notes, ETH, Zurich. With permission.)

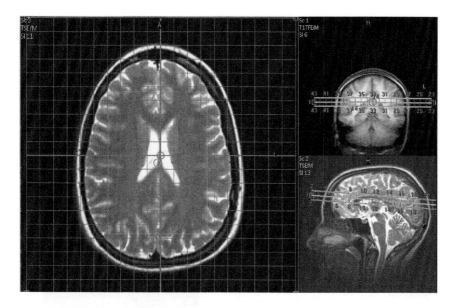

Figure 10.10. 2D-CSI from the entire brain in the oblique-axial plane at the location of the anterior–posterior commissure. (Reproduced from GyroTools, Inc., Lecture Notes, ETH, Zurich. With permission.)

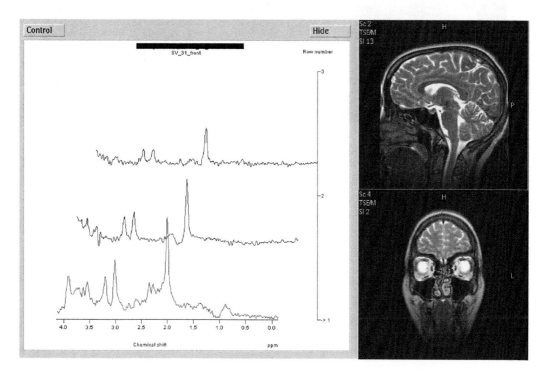

Figure 10.11. Typical single-voxel ^{1}H spectrum from the frontal cortex of a human volunteer. (Reproduced from GyroTools, Inc., Lecture Notes, ETH, Zurich. With permission.)

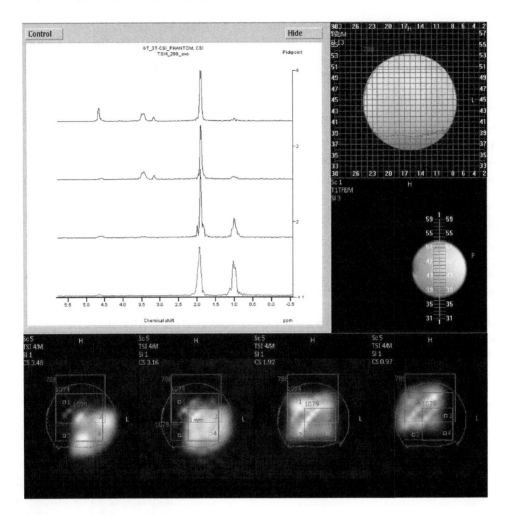

Figure 10.12. Spectroscopic imaging of the different moieties in the spectroscopy phantom. Spectral peaks are filtered, fitted, and then edited to allow their subsequent spectroscopic imaging. (Reproduced from GyroTools, Inc., Lecture Notes, ETH, Zurich. With permission.)

10.4.4   Artifacts in Spectroscopy

There are numerous typical artifacts encountered in spectroscopy. The most common artifacts are listed below.

10.4.4.1   Delayed Acquisition of FIDs

In both CSI and DRESS imaging there is a short delay (t_D) between acquisition and detection, during which the gradients are on, to allow refocusing the population of the excited spins or phase encoding them. Problems associated with such acquisitions include the following.

10.4.4.2   Short T_2 Moieties Leading to Signal Loss

Since there exist multiple moieties within the selected BW that resonate with slightly different frequencies compared to the center frequency of the NMR pulse, each of these dephases by a different amount during t_D, giving rise to a first-order (linear) phase variation in the acquired spectrum. For a moiety that

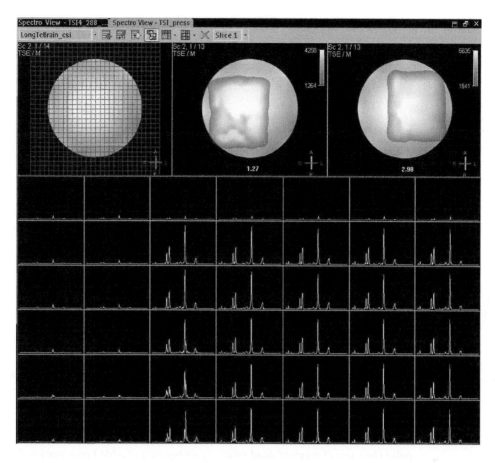

Figure 10.13. Direct spectroscopic imaging of separate moieties contained in the spectroscopy phantom. (Reproduced from GyroTools, Inc., Lecture Notes, ETH, Zurich. With permission.)

is chemically shifted by δ Hz from the center frequency of the NMR pulse, the phase difference is

$$\phi(\delta) = 2\pi.\delta.t_D \tag{10.35}$$

Additional contribution to first-order (and zero-order) variation in the spectrum is caused by other sources, such as receiver filters, as described below. In general, the phase of the sampled NMR signal does not match the phase of the receiver. If we denote such phase difference as φ, the detected FID signal takes the form

$$f'(t) = \sum_{j=1}^{k} e^{i(\phi+\omega_j t)} e^{-\lambda_j t} \tag{10.36}$$

and its Fourier transform becomes

$$F'(\omega) = (\cos \varphi + i\sin \varphi) \cdot [A\,(\omega) + iD(\omega)] \tag{10.37}$$

Therefore, the real and imaginary channels in the computer will contain mixtures of the absorption and dispersion line shapes (Figure 10.14).

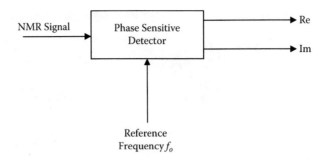

Figure 10.14. Phase-sensitive detection.

$$\text{Re } [F'(\omega)] = \cos \varphi \cdot A(\omega) - \sin \varphi \cdot D(\omega) \tag{10.38}$$

$$\text{Im } [F'(\omega)] = \sin \phi \cdot A(\omega) - \cos \phi \cdot D(\omega) \tag{10.39}$$

Thus,

$$A(\omega) = \cos\phi \cdot \text{Re}[F'(\omega)] + \sin\phi \cdot \text{Im}[F'(\omega)] \tag{10.40}$$

Such a process is known as phase correction. Typically, it is carried out by manual, interactive adjustment of φ, until a pure absorption (or dispersion) line shape is obtained.

Furthermore, such a phase correction process may be a function of frequency. In many cases, sampling of the FID does not begin immediately after the RF pulse (because of receiver dead times, t_D, or because of added propagation delays of the audio frequency filters used to remove high-frequency noise). In such a case the FID is described by

$$f(t) = \sum_{j=1}^{k} e^{i[\omega_j (t+t_o)+\phi]} e^{-\lambda_j t} \tag{10.41}$$

where $\varphi_{tot} = \varphi + \omega_j \, t_o$. Substitution of this equation in 10.41 (and considering a homogeneous field and single T_2 spin response) yields

$$f(t) = \sum_{j=1}^{k} e^{i\phi_{tot}} e^{i\omega_j t} e^{-\lambda t} \tag{10.42}$$

In effect, proper phase correction of the spectrum requires both zero-order ($\varphi = $ constant) and first-order ($\varphi \approx \omega_j t$) correction terms. On these, the $\phi(\sigma)$ term needs to be added to yield the compound phase.

Loss of data during t_D leads to significant baseline artifacts. In effect, if the first few points in the acquisition of data are missed, this can be equivalent to neglecting the DC term in the Fourier series representation of either the FID or the spatially encoded data array. Setting the missing data to zero is equivalent to multiplying the true complete time domain signal by a boxcar function $g(t)$:

$$s_{trunc} = s(t) \cdot g(t) \tag{10.43}$$

In the Fourier domain,

$$S_{trunc}(\omega) = S(\omega) * G(\omega) \tag{10.44}$$

where

$$S(\omega) = \sum_{j=1}^{k} \frac{\left[\dfrac{1}{T_2}\right]}{\left(\dfrac{1}{T_2}\right)^2 + (\omega_j - \omega)^2} \tag{10.45}$$

and

$$G(\omega) = \int_{0}^{\infty} g(t) e^{-j\omega t} dt \tag{10.46}$$

$$= \int_{t_D}^{T} g(t) \cdot e^{-j\omega t} dt \tag{10.47}$$

$$= \int_{t_D}^{T} e^{-j\omega t} dt \tag{10.48}$$

$$= \left(\frac{T - t_D}{2}\right) \cdot e^{-j\frac{\omega T}{2}} \sin c\left[\frac{\omega(T - t_D)}{2}\right] \tag{10.49}$$

Effectively, the observed spectrum is a convolution of the true spectrum with the sinc function. This leads to the depression of the baseline around each peak and propagates the rippling pattern along the baseline. When the spectral peaks are very close to each other, or when they overlap, such wiggles may interfere, distorting the spectra. This can have adverse effects on the absolute quantification of spectra.

Numerous solutions to this problem have been proposed in the literature, including least-squares fitting of the time domain data, use of the maximum likelihood estimation theory, use of maximum entropy reconstruction of spectra, and use of the statistical spectral estimation theory. A simple yet powerful method to solve such a problem was proposed by Allman et al. (1988). In this method, copies of the sinc function corresponding to the FT of a step function of width T_D are generated and shifted to the frequencies of the major peaks in the spectrum, and scaled by their amplitudes.

Spectral correction in the Fourier domain involves taking the Fourier transform of $g_c(t)$, scaling and shifting it to all major peaks of the spectrum, and adding the results.

10.4.5   Fourier Bleed

This effect arises from the use of discrete FT for spatial reconstruction (CSI, RFZ) when the sample contains higher spatial frequencies than those encoded by the localization technique. If we have a sample of n time domain sets of data, spatially encoded in the xy dimension, they are then convolved by (Bottomley et al. 1989b)

$$\Phi(x) = \frac{\sin \pi x}{n} \sin\left(\frac{\pi x}{n}\right) \tag{10.50}$$

due to the application of the discrete FT. If the signal source is located at exactly $x = 0$, then the discrete FT provides an exact reconstruction since $\Phi(x \neq 0) = 0$. If the source falls somewhere between the spatial points, then an artifactual signal is deposited in adjacent points.

10.4.6 Spectral Filtering

Often the spectra (in the form of either an FID or a spatially encoded signal) are multiplied with a decaying exponential filter function to enhance their SNR. Maximum SNR enhancement is achieved using a matched filter, a function that shares the same time dependence as the signal. Mathematically, in accordance to the Fourier transformation properties of a product:

$$FT[f(t) \cdot g(t)] = FT[f(t)]*FT[g(t)] \tag{10.51}$$

$$= F(\omega)*G(\omega) \tag{10.52}$$

where $G(\omega)$ represents the transfer function (frequency response) of the filter. Clearly, the convolution process leads to a degradation of the spectral resolution, but also to an SNR increase. In typical spectroscopic applications, the broadening filter is of the order of 3 Hz for ^{1}H spectra, and 6–8 Hz for ^{31}P spectra.

Selected Readings

1. Bottomley PA. State of the Art. Human In Vivo NMR Spectroscopy in Diagnostic Medicine: Clinical Tool or Research Probe? *Radiology* 1989a; 170:1–15.
2. Bottomley PA. MR Spectroscopy of the Human Heart: The Status and the Challenges. *Radiology* 1994; 191(3):593–612.
3. Bottomley, PA, Hardy CJ, Roemer PB, Weiss RG. Problems and Expediencies in Human 31P Spectroscopy. *NMR in Biomedicine* 1989; 2(5–6): 284–289.

11 Advanced Imaging Techniques: Parallel Imaging

11.1 Introduction to Parallel Imaging

In accordance to prior descriptions, Fourier encoding and legicographic, rectilinear k-space sampling have dominated magnetic resonance imaging (MRI) since Professor R. Ernst's (1996) classic work on pulsed nuclear magnetic resonance (NMR). Despite limited, but significant, attempts to devise alternative k-space sampling/acquisition schemes (spiral and twisted projection imaging, echo planar imaging (EPI)) to minimize imaging times, radio frequency (RF) excitation has been invariably combined and used with Fourier encoding to achieve image construction (Grant and Harris 1996; Stark and Bradley 1999).

Fundamentally, parallel imaging (PI) attempts to reduce total imaging time, based on alternative RF encoding schemes, a principle adopted and widely used in chemical shift imaging (CSI) spectroscopy, where localization is achieved based on the sensitivity profile of the RF coil employed.

This chapter introduces the reader to the fundamentals of PI, and discusses briefly techniques widely used nowadays (including sensitivity acquisition at spatial harmonics (SMASH) (Sodickson and Manning 1997), sensitivity encoding (SENSE) (Pruessmann et al. 1999), and generalized autocalibrating partially parallel acquisition (GRAPPA) (Griswold et al. 2002)). Recently developed techniques on transmit arrays are also presented.

11.2 Parallel Imaging Fundamentals

11.2.1 Principles of PI

Shortly before the introduction of the NMR phased arrays (Section 7.5.3) by Dr. Peter Roemer et al. (1990), Carlson (1987) demonstrated the possibility of spatial encoding based upon the spatial field response characteristic of an RF coil. Despite the little importance and attention such work received, the idea was also independently conceptualized and formulated by Peter Roemer in his 1990 publication, when alternative reconstruction methods (to sum-of-squares image) were considered (Figure 7.18).

Within approximately one decade from such initial work, SMASH (1997), SENSE (1999), and GRAPPA (2002) were developed as manifestations of PI, with use of phased array coils. All such methods propose the acceleration of imaging acquisition through reductions of the number of phased encoding steps, and the concurrent use (and a priori knowledge) of the array coil spatial field response characteristics to synthesize/reconstruct the missing k-space lines (SMASH, GRAPPA) or eliminate aliasing (artifacts) in the image domain (SENSE). In such a way, only a limited number of phase encoding steps is executed (leading to accelerated imaging acquisitions) with the spatial resolution (dependent on $k_{x,max}$, $k_{y,max}$; See Equations 5.9 and 5.10 and image contrast remaining unaltered.

The term *parallel imaging* (PI) arose from the concurrent and parallel acquisition of data constructing the various subparts of the final image, by each of the array coils. Modern clinical systems employ 8–16 receiver channels (and are thus capable of using phased arrays with as many as 8–16 coil elements), while the new generation of scanners can allow as many as 32 channels (Blamire 2008). Research systems can of course support multiples of tens to hundreds of channels, with the limiting factor being the prohibitively high cost of receiver units (Blamire 2008; Schmitt et al. 2007).

Effectively, the existence of multiple, independent, parallel receive RF channels, in association with devised schemes of reduced spatial encoding, can allow PI at reduced imaging times.

Over the years, a number of variants of such techniques have been developed (Blaimer et al. 2004), including, but not limited to, AUTO-SMASH (Jakob et al. 1998), VD-AUTO-SMASH (Heidemann et al. 2001), Generalized SMASH (Bydder et al. 2002), MSENSE (Wang et al. 2001), partially parallel imaging with localized sensitivities (PILS) (Griswold et al. 2000), and sensitivity profiles from an array of coils for encoding and reconstruction in parallel (SPACE RIP) (Kyriacos et al. 2000). The basic technical concepts of the three major techniques are presented, using PILS as an example for introduction of SMASH.

11.2.2   SENSE, SMASH, and GRAPPA

The three major PI techniques can be classified into image space (SENSE) and k-space (SMASH, GRAPPA) techniques.

If Cartesian-based sampling is considered (spin–warp imaging), undersampling k-space acquisition (by omission of alternating k-space lines, while maintaining maximal $k_{x,max}$, $k_{y,max}$ values), leads to aliasing (along the phase encoding direction), as shown in Figure 11.1.

The degree of k-space undersampling (2×, 4×, etc.) accelerates acquisition by corresponding times (2×, 4×, etc.). This undersampling is generalized in the form of a reduction factor term, known as the acceleration factor R.

SENSE (Pruessmann et al. 1999) is a manifestation of an "unfolding" algorithm in the case where subsampling and phased arrays are employed for accelerated image acquisition and reconstruction. Each image pixel (in the aliased, reduced field of view (FOV)) in each of the images reconstructed from individual coils contains information from multiple (equal to R) other pixels (equidistant to its location), within the FOV of the unaliased (fully sampled) image. However, each of these pixels is weighted in intensity, in the original (unaliased) image, by the B_1 field sensitivity pattern S_k of the array, comprised of k-coils.

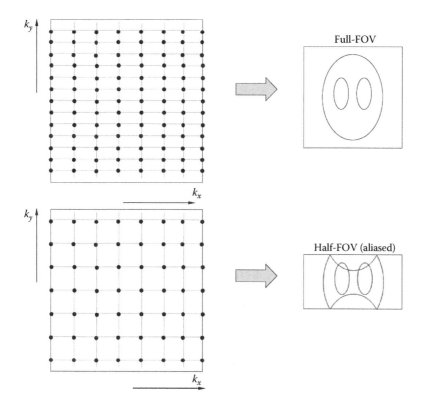

Figure 11.1. Reduced FOV and spatial aliasing as a result of k-space undersampling.

Modification of the generalized imaging Equation 5.7 to accommodate SENSE encoding leads to the k-space data formulation in accordance to

$$s[k_r(t)] = \sum_{n=0}^{N-1} S_{k,n} e^{-i\omega_n t} \rho(x,y).e^{-i.2\pi.k_r.r} \tag{11.1}$$

where $S_{k,n}$ is the sensitivity pattern for coil k in the phased array (comprised of N_c coil elements), weighted in phase by the term $e^{-i\omega nt}$.

The equivalent formulation in image space for coil element k, $I_k(x, y)$, is (Blaimer et al. 2004)

$$I_k(x,y) = \sum_{n=1}^{N} S_{k,n} \rho(x,y_n) \tag{11.2}$$

Matrix formulation of Equation 11.2 leads to

$$I = S.\rho \tag{11.3}$$

where I represents the complex image vector (with a dimension of $N_c \times 1$), S the complex sensitivity matrix for each coil at the aliased locations R (with a dimension of $N_c \times R$), and ρ the proton density function comprised of values for the R pixels of the full-FOV (unaliased) image.

Inversion of Equation 11.3 results in

$$\rho = (S^H S)^{-1} S^H . I \tag{11.4}$$

where H represents the Hermitian form of matrix S. Two critical issues that pertain to Equation 11.4 also account for finite (yet nonexistent in practical applications) noise correlations between the coils in the array (Harpen 1990; Constantinides et al. 1997), and incorporate knowledge of the phased weighted factors associated with S. The latter is often formulated as a geometric factor (G). While SENSE can implement PI without specific geometrical arrangement of the k-coils in the array (in comparison to other techniques, such as SMASH), both the geometric and acceleration factors ultimately limit SNR is SENSE in accordance to

$$SNR_{SENSE} = \frac{SNR_{unaliased}}{G.\sqrt{R}} \tag{11.5}$$

Such reductions (associated with faster image acquisitions) are offset and counterbalanced by the inherently higher SNR performance of the phased array.

SENSE is a method widely established nowadays with direct implementation by most imaging companies (in various forms), including SENSE (Philips), mSENSE (Siemens), ASSET (General Electric), and SPEEDER (Toshiba) (Blaimer et al. 2004). It has become (at acceleration factors of 2–4) part of clinical routine practice. Recent implementations of efficient SENSE algorithms (TSENSE) have led to ultra-fast, 2D cardiac imaging at sub-breath-hold acquisitions (Kellman 2001a, 2001b).

Some of the major limitations and disadvantages of SENSE (Noll) include the increased data requirements and complicated image reconstruction.

11.2.3   SMASH and GRAPPA

SMASH operates in k-space and attempts to reconstruct missing k-space values (due to missing phase encoding steps). In a similar fashion to SENSE, SMASH employs the coil sensitivity patterns $S_{k,n}$ of the array coils. In effect, spatial sensitivity patterns are weighted by linear weights n_k^m to estimate the missing spatial harmonics, in accordance to (Blaimer et al. 2004)

$$S_m^{comp} = \sum_{k-1}^{N_c} n_k^m S_k \cong e^{im\Delta k_y y} \tag{11.6}$$

Linear weights can be computed by fitting the actual array coil sensitivity patterns to the desired spatial harmonic function $e^{im\Delta k_y y}$.

Reducing Equation 11.6 for the elicited signal along the phase encoding direction only results in

$$s(k_y) = \int \rho(y).e^{-ik_y y} \, dy \tag{11.7}$$

Therefore, weighting the elicited signal of Equation 11.7 by the coil sensitivity function and linear weights of Equation 11.6 leads to the missing and shifted lines of k-space:

$$\sum_{k-1}^{N_c} n_k^m s(k_y) = \int \rho(y). \sum_{k-1}^{N_c} n_k^m S_k . e^{-ik_y y} \, dy = s^{comp}(k + m\Delta k_y) \tag{11.8}$$

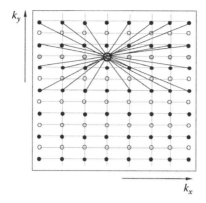

Figure 11.2. Schematic representation of missing k-space data using GRAPPA. The "spider's sampling net"-estimation of the missing data at the jth data location uses information from acquired neighboring data samples (only four acquired data lines are shown for clarity). Filled circles represent (black) acquired data points and (gray) data samples to be estimated; empty circles represent missing data points.

Equation 11.8 represents the SMASH relation. Clearly, a major limitation for this technique is its dependence to coil configurations along the phase encoding direction for accurate estimation of missing k-space data and reconstruction.

GRAPPA, although similar to SMASH, tries to reconstruct/estimate the missing k-space data by weighting the elicited signal from each coil $s_k(k_y)$ and not from the composite k-space signal $s(k_y)$. The method also uses multiple k-space lines from all coils to estimate missing k-space data from a single coil matrix (Figure 11.2). In this respect, Equation 11.8 thus becomes

$$\sum_{k-1}^{N_c} n_k^m s(k_y) = s_k^{comp}(k + m\Delta k_y) \tag{11.9}$$

GRAPPA generates single-coil images (that contain estimated missing data). This is highly advantageous since the composite image can be reconstructed using the conventional sum-of-squares reconstruction, alleviating and eliminating phase problems (as a result of the magnitude operation), and yielding higher image SNR.

Similar to SENSE, GRAPPA does not necessitate specific geometric placements of array coils. It has been implemented in the clinical setting and is readily available for use nowadays. Its performance is similar to that of SENSE; however, in the cases where accurate coil sensitivity representation is possible, SENSE is optimal in SNR response.

11.2.4 Coil Sensitivity Determination and Autocalibration Procedures

Coil sensitivity characteristics are critical for PI techniques. While a number of imaging companies have incorporated (in their software) capabilities (from within the implemented sum-of-squares algorithm) to turn on control variables to generate composite images and individual coil sensitivity maps, alternative techniques (such as a low-resolution 3D Fourier-encoded imaging) are often conducted prior to PI encoded acquisitions.

Implementation of autocalibration procedures (first introduced by Jakob et al. (1998)) provides the most sensible alternative, where such coil estimations are completed in a prescan fashion before (or after) encoding, thereby eliminating any inconsistencies due to patient motion or breathing variability, or coil loading mismatches (especially in the cases of use of flexible phased arrays).

11.3   Transmit Phased Arrays

A typical discussion on arrays would conceptually start with the topic of transmit arrays. However, this unorthodox presentation pattern has been endorsed in this case, primarily because it best matches prior and current efforts in the field. Receive phased arrays were introduced and developed first. In addition to research developments, the corporate imaging world needed to make significant hardware (and software) adjustments to existing clinical (or research) systems to accommodate changes.

In a similar fashion, the emergence of transmit phased arrays followed technological changes. Their development is still in the early stages, and—similar to receive arrays—their incorporation to clinical and research systems will also require minor and major hardware redesign (amplifiers, multitransmit channel capability and synchronization, transmit coil design and interface, etc.). While transmission and reception can be potentially implemented from the same coil array, the capability to decouple array transmission and reception will be implemented.

The concept of transmit arrays finds its origin in the conventional method of transmission-reception in modern systems; a highly homogeneous birdcage (body) coil is used to excite spins, and the same coil, or a reception coil, or a coil array, is used to detect the elicited signal. Often, the homogeneity of the excitation profile falls short of expected standards or values, leading to variability in sensitivity and image contrast, effects that can prove detrimental (for example, when differential saturation is desired in T_1-weighted imaging). Interactions of the excitation RF with the tissue can also lead to an even further heterogeneous excitation, becoming worse at higher field strengths, where the excitation wavelength becomes comparable to the spatial dimensions of the imaged objects, leading to standing wave patterns (Blamire 2008). To date, spectral-spatial or adiabatic RF pulses have been used to alleviate inhomogeneous B_1 field excitation issues; however, they are associated with long execution times and an increased complexity in their implementation.

The presence of multiple, independent transmit RF channels and transmit arrays has been recently implemented to allow tailored RF transmission, a term known as RF (or B_1) shimming, leading to a more uniform excitation profile. In effect, the composite excitation profile $M(r_m)$ is the summation of the spatial excitation profiles of each of the k-array coils, $m(r_m)$, weighted by transmission weight factors $S_k(r_m)$, in accordance with the geometrical array placements around the imaged subject.

$$M(r_m) = \sum_k S_k(r_m) m_k(r_m) \qquad (11.10)$$

For a given desired excitation profile $M(r_m)$, the shape of the individual RF pulses necessary for each array coil excitation can be computed via inverse estimation of the Fourier transform of Equation 11.10, solving for $m_k(r_m)$.

Reception can progress in accordance with conventional reception schemes, or further to implementation of PI techniques and methodologies, as discussed previously.

Problem Sets

Chapter 1: Fourier Transformations

1. Given a 2D continuous signal $g(x, y) = (x + \alpha y)^2$, evaluate $g(x, y).\delta(x - 1, y - 2)$ and $g(x, y)*\delta(x - 1, y - 2)$.

2. Consider two 2D continuous signals $f(x, y)$, $g(x, y)$ that are separable, i.e., $f(x, y) = f_1(x).f_2(y)$ and $g(x, y) = g_1(x).g_2(y)$.

   a. Show that their convolution is also separable

   b. Express the convolution in terms of $f_1(x)$, $f_2(y)$, $g_1(x)$, $g_2(y)$

3. Two linear shift invariant (LSI) systems are connected in serial or parallel cascade. Prove that the resulting LSI system is also LSI and determine the point spread function (PSF) and point spread sequence (PSS) of the overall system (consider both the continuous and discrete cases).

4. Calculate the Fourier transformation of a rectangular function in one and two dimensions. (Optional) Repeat the calculation for a sinc function with period T, and a Gaussian function with amplitude A and standard deviation σ.

5. a. Prove the shifting property of the Fourier transformation of a discrete function, such that

   $$FT\{g[m - m_o, n - n_o] = G(\omega_x, \omega_y) \cdot e^{-i(\omega_x m_o + \omega_y n_o)} \qquad \text{(Q 5.1)}$$

   where m_o, n_o are integer constants and $i = \sqrt{-1}$.

   b. The Fourier transformation $X(\omega_x, \omega_y)$ of a discrete function $x(m, n)$ is given by

   $$X(\omega_x, \omega_y) = 1 + 2e^{i\omega_x} + 2e^{-i\omega_x} + 4e^{-i\omega_x} e^{-i\omega_y} \qquad \text{(Q 5.2)}$$

   Estimate the discrete series $x(m, n)$ and plot it.

6. a. Define the Fourier transformation and inverse transformation of 2D continuous functions.

   b. Prove the scaling property in continuous space:

   $$FT\{I_1(ax,\beta y)\} = \frac{1}{|\alpha\beta|} I_1\left(\frac{\omega_x}{a}, \frac{\omega_y}{\beta}\right) \qquad (Q\ 6.1)$$

   where α and β are constants.

   c. How will a typical MR image I change if the image is filtered using a Gaussian filter Φ such that $F = I*\Phi$?

   d. A typical image is represented by matrix I. Compute the resulting image after application of a mean filter Φ:

   $$I = \begin{bmatrix} -2 & 0 & 1 \\ 3 & 4 & 7 \\ 4 & -2 & 5 \end{bmatrix}, \Phi = \begin{bmatrix} 1 & 0 \\ 0 & 1 \end{bmatrix} \qquad (Q\ 6.2)$$

7. If

$$g(m,n) = \frac{m - \dfrac{2}{n}}{(m-3n)^2} \qquad (Q\ 7.1)$$

Compute the following:

   a. $g(m, n).\delta(m, n-3)$ \hfill (Q 7.2)

   b. $g(m, n)**\delta(m-3, n-1)$

   c. Write the Fourier transformation of a discrete image $x(m, n)$ in two dimensions such that

   $$x(m,n), \quad -\infty,...,-1,0,1,...+\infty \qquad (Q\ 7.3)$$

   Compute and sketch the discrete function $x(m, n)$, given that $X(\omega_x, \omega_y)$ is given by

   $$X(\omega_x,\omega_y) = 3 + e^{-j\omega_x} + j8\sin\omega_y + 2e^{-j\omega_x}e^{-j\omega_y} \qquad (Q\ 7.4)$$

   where $j = \sqrt{-1}$.

8. If

$$g(x,y) = \frac{1}{(x-3y)^3} \qquad (Q\ 8.1)$$

Calculate the following:

   a. $g(x, y)\cdot\delta(x, y-1)$

b. $g(x, y)**\delta(x - 3, y - 1)$

c. Prove the convolution property of the Fourier transformation, that is, for two two-dimensional images $I_1(x, y)$ and $I_2(x, y)$:

$$FT\{I_1(x, y)**I_2(x, y)\} = I_1(\omega_x, \omega_y)I_2(\omega_x, \omega_y) \qquad \text{(Q 8.2)}$$

where $I_1(\omega_x, \omega_y), I_2(\omega_x, \omega_y)$ represent the mathematical transformations of the 2D images $I_1(x, y)$ and $I_2(x, y)$.

Chapters 2–7 and 9: Magnetic Resonance

1. The following equation,

$$M_z(t) = M_o(1 - e^{-t/T_1}) + M_o \cos\alpha . e^{-t/T_1} \qquad \text{(Q 1.1)}$$

gives the components of M for an α pulse, when the system is in equilibrium, just before the α pulse. Suppose that the sample is now irradiated with a train of α pulses, separated by a time TR. The equilibrium condition is true when TR is long compared to T_1, and it can be assumed that M_z just before the pulse is equal to M_o. Derive a more general formula for $M_z(t)$. It can be assumed that the transverse magnetization has completely dephased before each RF pulse, i.e., $M_{xy}(\text{TR}) = 0$. (Hint: In this more general formula, M_o will be replaced with the steady-state value of the longitudinal magnetization. Define M_z after the $(n + 1)$th pulse to be M_z^{n+1}, and M_z after the nth pulse to be M_z^n. Relate these two quantities with an equation. Derive another (very simple) equation from the steady-state condition.) (Courtesy of Professor E. McVeigh.)

2. What repetition time (TR) should be used to generate the maximum signal difference between the two tissues whose T_1 values are T_1^a and T_b^1? What TR value will give the maximum signal difference to the noise ratio? (Note: Assume that TE is zero, and that all transverse magnetization has decayed before each $\pi/2$ pulse.)

3. a. What is the slice thickness (defined as full width at half maximum) if a slice selection gradient amplitude of 1 Gauss/cm is used with a Gaussian-shaped RF that has a shape given by $A(t) = A_o e^{\{-t^2/\sigma^2\}}$, where $\sigma = 1$ ms?

    b. What is the slice thickness if the gradient amplitude is cut in half? Suppose the shape of the RF pulse is changed so that σ is reduced by a factor of two.

    c. What is the new slice thickness (with a 1 Gauss/cm gradient)?

    d. What else is affected by this change?

    e. Derive the phase $\varphi(z)$ in the z direction after a slice selection with a constant z grad (assume infinite slew rates and no refocusing lobe).

4. Recall that the imaging equation was given by the following:

$$s(n,m) = e^{-(nT+t_y)/T_2} \iint \rho(x,y) e^{-i\gamma(nTG_x x + mt_y \Delta G_y y)} dx dy \qquad \text{(Q 4.1)}$$

$$0 \le n \le N \qquad \text{(Q 4.2)}$$

$$-M/2 + 1 \le m \le M/2 \qquad \text{(Q 4.3)}$$

What is the effect of the $e^{-(nT+t_y)/T_2}$ term on the image? Explain.

5. (Contributed by Dr. B. Bolster.) Suppose a point object is moving along the x axis with the following trajectory: $x(t) = x_0(t) + vt$.

   a. Calculate the phase shift that is induced in the transverse magnetization after the application of the gradient waveform shown below (Figure Q5.1):

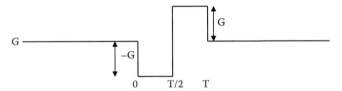

Figure Q5.1. Gradient waveform.

   b. Calculate the phase shift induced by the waveform shown below (Figure Q5.2):

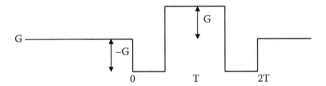

Figure Q5.2. Revised gradient waveform for flow compensation.

   This is called a flow compensation pulse.

   c. Repeat the calculations for the above two waveforms for a trajectory given by $x(t) = x_o(t) + vt + 1/2\alpha t^2$.

   d. What gradient waveform could be used as an acceleration compensation pulse?

   e. Is it possible to design a gradient waveform that will produce phase shifts that are independent of acceleration and dependent on the velocity?

6. (Questions 6–8 contributed by Professor E. McVeigh.) Suppose that a slice has been selected through a $5 \times 5 \times 5$ cm³ cube. The goal is to produce a projection of this slice with the application of a readout gradient. The pulse sequence in Figure Q6.1 is used to produce the profile.

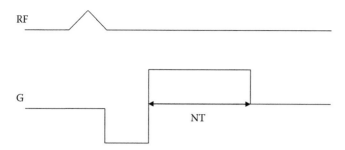

RF

G

NT

Figure Q6.1. Pulse sequence diagram.

If the gradient strength G = 1 Gauss/cm and N = 256 sample points are taken in the total sampling time NT = 10 ms:

a. What is the spatial extent of the profile? (This is referred to as the field of view (FOV).) What is the bandwidth of the received signal?

b. How many pixels does the profile of the object span?

c. Suppose we set G = 0.5 Gauss/cm, N = 256, and NT = 20 ms. How does this affect a and b?

d. Repeat a–c for the following pulse sequence (Figure Q6.2):

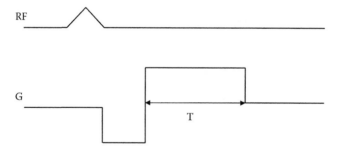

RF

G

T

Figure Q6.2. Revised pulse sequence diagram.

7. Show that the inverse Fourier transform of $s(t)$ in the imaging equation is the projection of $\rho(x, y)$ onto the z axis convolved with a Lorentzian.

8. Suppose an object is wished to be imaged in all three dimensions, that is, its final image is going to be an $N \times N \times N$ array of voxels with isotropic dimensions. However, the ability to produce RF pulses is limited to short bursts of unmodulated RF, and therefore slice selection cannot be used.

a. What is a proposed method for imaging the object? (Hint: Revisit the phase encoding section to extend the presented ideas.)

b. What would the equation for the signal be?

c. How can the image be reconstructed?

9. (Contributed by Dr. S. Reeder.)

   a. Determine an expression for the longitudinal magnetization immediately before the $(n + 1)$th RF pulse from a simple train of nonselective degree pulses separated by TR. It may be assumed that TR $\gg T_2$ (but not TR $\gg T_1$).

   b. Calculate the steady-state longitudinal magnetization assuming that the magnetization before the nth RF pulse equals the magnetization before the $(n + 1)$th RF pulse. What is the transverse magnetization immediately after the $(n + 1)$th RF pulse?

   c. Using the solution for the steady-state signal determine the optimum flip angle that maximizes the signal. This angle is known as the Ernst angle. Also describe in words why it is important to choose the tip angle wisely.

   d. Write an expression that relates the SNR as a function of TR normalized for constant scan time. Use $T_1 = 1000$ ms, and the Ernst angle for each TR. What is a preventing factor from reaching the optimum?

   e. Without making any assumption, calculate a closed-form solution for the longitudinal magnetization immediately before an RF pulse, for an arbitrary number of RF pulses $(n + 1)$. What happens as n tends to infinity?

   f. Last, explain how you would determine the tip angle that maximizes the total accumulated signal over n RF pulses. How does the optimum tip angle compare to the Ernst angle?

10. Modify the imaging equation to include the following:

    a. The effect of a fat-water chemical shift. Why does a chemical shift only occur in the readout direction for spin–warp imaging?

    b. The effect of T_2. What is the spatial resolution in the readout direction for a specific T_2 value? What is the highest spatial resolution achievable with a gradient set of specific performance features and a specific T_2 value?

    c. The effect of motion. What is the blurring in the readout direction for a specific velocity?

11. What would be the single RF pulse necessary to excite two slices simultaneously?

12. Given a point object at origin:

    a. Write down the equation for the free induction decay (FID), given that the transverse magnetization decays at a rate of T_2.

    b. What is the expression of the complex frequency spectrum? What is the expression for the power spectrum (magnitude of the Fourier-transformed signal)?

c. Let $T_2 = 50$ ms. Obtain the full width at half maximum (FWHM) in Hz from the real part of the frequency spectrum and from the power spectrum. What is the effect of a magnitude operation on the FWHM?

d. What effect does this have on an image of a continuous object?

13. (Contributed by Dr. S. Reeder.)

a. Using the definition of the Fourier transform derive explicitly the time representation of a signal $s(t)$ such that

$$X(\omega) = \begin{cases} 1 & if \; |\omega| \le 12.56x10^3 \, rad/s \\ 0 & \text{otherwise} \end{cases} \qquad \text{(Q 13.1)}$$

b. Sketch $s(t)$ and label clearly the zero crossings and the maximum amplitude.

c. Show explicitly how you can modify the signal $s(t)$ to ensure frequency selection between 3 and 7 kHz. (Hint: Recall the shift theorem of the Fourier transform.)

d. What would be the effect of truncation of $s(t)$ in the frequency domain? Propose a scheme to alleviate expected problems from such an operation.

14. Recall the imaging equation:

$$s(n,m) = e^{-(nT+t_y)/T_2} \iint \rho(x,y) e^{-i\gamma(nTG_x x + mt_y \Delta G_y y)} dxdy \qquad \text{(Q 14.1)}$$

$$0 \le n < N \qquad \text{(Q 14.2)}$$

$$-M/2 + 1 \le m \le M/2 \qquad \text{(Q 14.3)}$$

a. Write down an expression for the imaging data from a point object that is at $y = 0$ and is undergoing sinusoidal motion in the z direction during the acquisition (neglecting T_2 decay).

b. What two effects will this motion have on the appearance of the final image?

c. Suppose that TR = 1 s and the frequency of the oscillation is 0.33 Hz. The amplitude of the oscillation is one pixel unit, and the point object starts on a pixel grid position for the first acquisition. Make a labeled sketch of the resulting image.

15. a. Describe the methodology for selection of an image slice in the plane X-Z in three-dimensional space (in oblique orientation) in the Cartesian reference system, with a plane rotation of 45°. Draw the pulse sequence diagram for spin–warp imaging. Consider that

• X grad = Frequency encoding

• Y grad = Phase encoding

- Z grad = Slice selection

Use diagrams where/if needed.

b. Starting from the imaging equation:

$$s(n,m) = e^{-\frac{(nT+t_y)}{T_2}} \iint \rho(x,y)e^{-i\gamma(G_x xnT + m\Delta G_y t_y y)}\, dx\, dy \qquad \text{(Q 15.1)}$$

If the imaged object (within the field of view) undergoes motion along the X axis according to the equation of linear displacement, compute mathematically and describe the effects of such a motion on the reconstructed MR image, where

$$x = x_o + vt \qquad \text{(Q 15.2)}$$

where x denotes the object's position in space (X axis), x_o its initial position, and v its linear velocity.

c. The following (Figure Q15.1) represent geometrically distorted MR images from a phantom with a linear Cartesian grid. Considering the imaging equation and the use of gradient coils, describe in a simple manner the reason for such distortion.

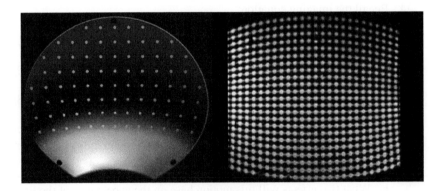

Figure Q15.1. Geometric distortion of Cartesian grid in MR images reconstructed with nonideal gradients.

d. Explain the physical principles for the relaxation times T_1, T_2. What is the biophysical meaning of the longitudinal and transverse relaxation constants in the molecular environment?

16. a. Explain the slice selection methodology for a transaxial slice along the Z axis in three-dimensional space of a Cartesian reference system, translated by 4 cm from the scanner isocenter. What is the slice thickness? Consider that

- Z grad = 2 Gauss/cm (slice gradient encoding)

- $\Delta t_{rf} = 3$ ms (sinc excitation radio frequency)

- $f_o = 63.89$ MHz (resonance frequency)

In your answer use diagrams if and where they may be needed.

b. Plot the *k*-space sampling trajectory that is defined by the following gradients $G_x(t)$ and $G_y(t)$ (Figure Q16.1) during an MRI experiment.

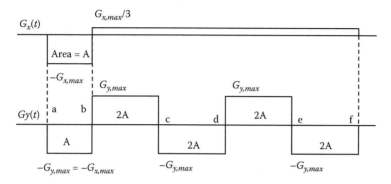

Figure Q16.1. Readout (G_x) and phase encoding gradients (G_y).

17. a. Transform the imaging equation through the use of the spatial frequencies k_x, k_y.

b. State the general mathematical equations that relate the spatial frequencies k_x, k_y with G_x, ΔG_y in the case of MRI. Rewrite these equations in such a manner such that the solutions relate the gradients G_x, ΔG_y with the spatial frequencies k_x, k_y.

18. a. State the mathematical definitions for image spatial resolution, contrast-to-noise ratio (CNR), and signal-to-noise ratio (SNR). In your answer use diagrams and equations if and where they are needed.

b. State the type of image slice for the following cardiac image (Figure Q18.1) from a human volunteer.

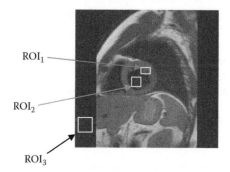

Figure Q18.1. Typical human cardiac MRI in a short-axis orientation.

c. In your effort to quantitatively assess the image quality, multiple measurements were conducted in various regions of interest (region of interest 1 (ROI_1), region of interest 2 (ROI_2), region of interest 3 (ROI_3)). The results are summarized in Table Q18.1:

Table Q18.1. Quantitative Estimates of Image Quality for the Image in Figure Q19.1

A/A	Signal Intensity (ROI_1)	Signal Intensity (ROI_2)	Signal Intensity (ROI_3)	Standard Deviation (σ) (ROI_1)	Standard Deviation (σ) (ROI_2)	Standard Deviation (σ) (ROI_3)
1	150	23	15	23	19	11
2	134	30	11	15	18	14
3	172	32	9	40	17	8
4	190	28	14	54	33	16

Calculate the:

   i. **Mean** SNR value of the image.

   ii. **Mean** CNR between the myocardium and the left ventricular blood chamber.

19. a. The closed-form equation for the total detected signal $s(t)$ in a brain MRI scan using a gradient–echo pulse sequence is given by

$$s(t) = M_o e^{-\frac{TE}{T_2}} \frac{\left(1 - e^{-\frac{TR}{T_1}}\right) \sin\alpha}{1 - e^{-\frac{TR}{T_1}} \cos\alpha} \qquad (Q\ 19.1)$$

Given that the tip angle $\alpha = 30°$, $M_o = 100$, $T_1 = 1000$ ms, $T_2 = 25$ ms, TR = 25 ms, and TE = 25 ms, calculate $s(t)$.

   b. At the end of the MR image acquisition the radiologist confirms that the image quality is poor for conclusive diagnosis. He suggests repetition of the exam using an intravenous contrast injection for the improvement of the signal-to-noise ratio. How (according to your view) does the contrast agent help to improve the overall detected image signal intensity? (Which parameters of the signal equation does it affect?)

   c. If the detected MR signal $s'(t)$ increases by 25% in b compared to the value calculated in a (that is, $s'(t) = 1.25s(t)$), how much have the affected parameter(s) changed after contrast injection?

   d. In the following diagram (Figure Q19.1) state the names of the major hardware components of the MRI scanner.

   e. Describe succinctly the basic function of each of the hardware components that you have listed above.

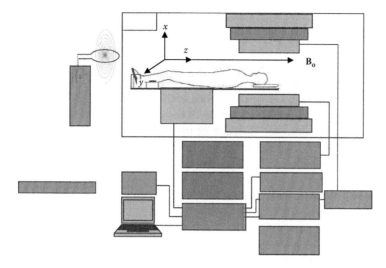

Figure Q19.1. Major hardware components for a modern MRI system.

    f. The basic principle of magnetic resonance based on quantum mechanics is based on Zeeman's phenomenon. Given that MRI is achieved by exciting ^1H nuclei, draw:

       i. The possible energy levels of the ^1H nucleus in the absence and presence of an external, static magnetic field B_0.

      ii. The nuclei distribution at the various energy levels showing the magnetization direction, and the equation of their population distribution in the presence of an external, static magnetic field B_0 (Boltzmann equation). Explain all the symbols. Is there a difference between the population numbers at the various energy levels?

     iii. State the equation of the energy difference of the energy levels ΔE based on the magnetic quantum number m_i.

     iv. Based on the above and considering the principles of magnetic resonance, list four ways with which you could increase the signal intensity of excited spins in an MRI experiment.

    Use diagrams if and where you think they are needed.

20. Nuclear magnetic resonance was discovered in the middle of the 20th century by Felix Bloch and Edward Purcell. Its use focused initially on chemical analysis applications until Paul Lauterbur introduced the concept of zeugmatography and gradient coils. The following excerpt is taken from one of Paul Lauterbur's classic publications (based on which he later received his Nobel Prize in Physiology or Medicine). Read it and summarize succinctly the meaning. In your response give special emphasis to the concept of zeugmatography.

True three-dimensional nuclear magnetic resonance images can be reconstructed by a two-stage filter back-projection algorithm. The mathematical analysis of the reconstruction technique has been presented previously (Lai and Lauterbur 1980). In a typical experiment, the zero-dimensional signal, obtained as the impulse response from the entire object in a homogeneous magnetic field, is converted to a one-dimensional representation of the spatial distribution of the NMR signals by the imposition of a magnetic field gradient. If the gradient is linear, each such representation corresponds to a projection resulting from integration of the signal over planes perpendicular to the gradient direction. Rotation of the gradient direction in a plane perpendicular to any axis generates a set of one-dimensional projections from which a two-dimensional image, as seen from a direction along that axis, may be reconstructed. Rotation of the gradient about another axis generates data from which another such image may be reconstructed. Any desired two-dimensional views may be obtained, without any mechanical motion of the object or the scanning apparatus, because linear magnetic field gradients combine vectorially, so that any desired gradient direction may be produced by appropriate adjustment of the currents in three sets of coils. From a set of two-dimensional views, or projections, a three-dimensional image may be obtained by a second stage of reconstruction, as shown in Figure Q20.1. Only three two-dimensional reconstructed projections are shown, for simplicity and clarity. Each has its normal lying in the yx plane, and from each, a strip of signal intensity may be chosen at a particular value of z and used to reconstruct a slice of the three-dimensional object. This procedure may be repeated to generate a complete three-dimensional array, which may be displayed as slice images with any orientation.

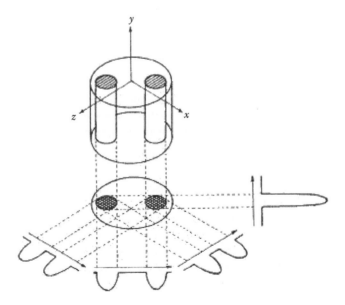

Figure Q20.1. Representative views used in a two-stage three-dimensional reconstruction. Each two-dimensional projection has been reconstructed from a number of one-dimensional projections. A reconstructed slice of the three-dimensional image is shown. The coordinate axes are those used in the experiments. (Reproduced from Lauterbur, P., *Nature,* 242(16), 190–191, 1973. With permission.)

a. Which other medical diagnostic technique is based on the principle of zeugmatography as the basis for imaging? Explain.

b. The process employed to generate a medical image is known as the reconstruction algorithm. Which is the primary reconstruction algorithm in a above, and what is the relevant theorem upon which image formation is based?

21. a. The classic description of spin–warp MRI employs the sequence of pulses as shown in Figure Q21.1.

    i. Explain the meaning of all symbols listed in the figure.

    ii. Describe briefly the use of pulses, gradients, and the way(s) such affect proton nuclei in ^{1}H MRI.

b. The use of small in amplitude and linearly spatially varying changes of magnetic fields (gradients) along the X and Y axes (Equations Q21.1, Q21.2) achieves nutational frequency encoding of spins and phase encoding within a selected (excited) slice. In an MRI center the physicist in charge verifies that the gradients G_x and G_y are imperfect. The gradients are shown to behave nonlinearly with spatial directions with a second-order degree dependence of gradients along the x or y axes (Equations Q21.3, Q21.4). Describe how the image quality is anticipated to change as the result of such imperfections.

$$B_x = k_1 . x \qquad\qquad\text{(Q 21.1)}$$

$$B_y = k_2 . y \qquad\qquad\text{(Q 21.2)}$$

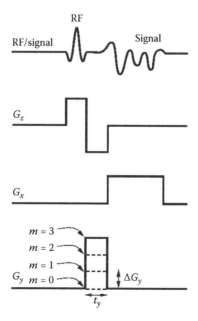

Figure Q21.1. Pulse sequence diagram for spin–warp imaging.

$$B_x = k_3 \cdot x^2 \qquad\qquad \text{(Q 21.3)}$$

$$B_y = k_4 \cdot y^2 \qquad\qquad \text{(Q 21.4)}$$

where k_1, k_2, k_3, and k_4 are constants.

c. Starting from the imaging equation:

$$s(n,m) = e^{-\frac{(nT+t_y)}{T_2}} \iint \rho(x,y) e^{-i\gamma(G_x xnT + m\Delta G_y t_y y)} \, dx\, dy \qquad \text{(Q 21.5)}$$

Compute mathematically and describe qualitatively and quantitatively the effect on the reconstructed MR image when the imaged object is a point object undergoing sinusoidal motion (within the field of view). Assume that the maximum amplitude of its motional trajectory is A, and that its period of oscillation is T/2 (at $t = 0$ the point object is at location (0, 0, 0) of the Cartesian coordinate system).

22. (Parts contributed by Professor E. McVeigh) The following pulse sequence is a modified spin–echo sequence. Using more than one 180° pulse (as shown below) multiple echoes can arise, occurring at every TE millisecond interval from the center of each echo. The maximum signal intensity of each echo exhibits a T_2 dependence. Such a pulse sequence is known as the fast spin–echo (FSE) pulse sequence. For such a sequence:

a. Sketch the gradients (G_x, G_y) that you must use to allow reconstruction of four separate images during each RF excitation. Each reconstructed image corresponds to different TE values. In your answer clearly label and explain all symbols used.

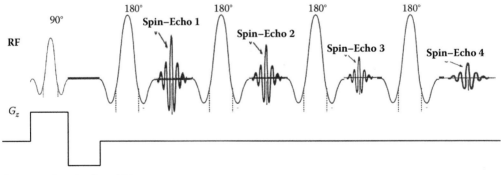

b. Sketch the k-space sampling trajectory showing clearly the correlation of the k_x-k_y diagram with the gradient diagrams in a above.

c. Modify the pulse sequence in a, above, such that the four echoes can be combined in the same signal acquisition matrix $s(n, m)$ based on the imaging equation, leading to a constructed image in 1/4 of the total acquisition time.

d. Sketch the modified k-space trajectory based on the modified pulse sequence of c, above. Use diagrams for your answers. Explain all symbols used when and where needed.

e. Which parameters would you envisage to change in c, above to generate T_2-weighted contrast images?

Chapters 8 and 10: Spectroscopy and Instrumentation

1. The quality factor Q of a radio frequency coil is 225 under unloaded conditions and 100 when loaded with a human body.

   a. What is the percentage noise voltage increase (as detected by the coil) as a result of its loading due to the presence of the human body (coil loading)?

   b. Assume that the signal-to-noise ratio (SNR) value in a specific MRI experiment of biological tissue is 64. Additionally, assume that the electronic noise of the systems (preamplifier, spectrometer, receivers, cabling) is negligible. What is the maximum attainable value of SNR if the coil noise is also negligible?

2. What would be the expected rise of SNR when the same MRI experiment is repeated in scanners of 1.5 and 4.7 T field strengths, assuming that the electronic noise of the systems (preamplifiers, spectrometer, receivers, cabling) is negligible, and that the coils, the image parameters of the pulse sequence employed in the two cases, are the same, and that:

   a. The human body (biological sample) is the dominant noise source.

   b. The RF coil constitutes the major noise source, in a similar fashion to a coil used for NMR microscopy?

3. Derive an expression for the axial magnetic field produced by a single turn square and cylindrical coils of size or radius a, carrying a current I, in free space, based on the Biot–Savart law of electromagnetism.

   a. An RF power amplifier delivers a 2 kW root-mean-squared RF pulse of duration 1 ms at a frequency of 64 MHz to a circular coil with a 20 cm diameter, matched to 50 ohm. Calculate the maximum B_1 field (in Tesla) at the center of the coil. Determine whether such a value complies with Food and Drug Administration (FDA) limits for RF power deposition.

   b. What is the B_1 field at a depth of 20 cm into the coil at its central axis?

   c. The unloaded Q factor is 200. The loaded is 100. What fraction of the transmitted power is deposited into the sample?

4. Sketch the proton spectrum of methanol, ethane, and acetone. Clearly mark your diagram to indicate chemical shifts, spin couplings, and their nature.

Multiple Choice Questions*

Section I

Select the best answer(s).

1. The transverse relaxation time T_2

    a. Is affected by the magnetic field homogeneity

    b. Is always shorter than or equal to T_1

    c. Affects the echo amplitude in a spin–echo experiment

    d. All of the above

    e. b and c

2. Some of the principle advantages of MRI over other imaging modalities include the following:

    a. Its ability to produce high-contrast images of soft tissues

    b. Its ability to image dynamic processes in real time

    c. Its ability to produce images in arbitrary planes

    d. All of the above

    e. a and c

3. Suppose a rectangular, nonselective RF pulse elicits a signal from a sample. Also, suppose that you know that this signal is 50% of the maximum obtainable. You suspect that the tip angle can be adjusted to obtain the maximum signal. You should

    a. Double the RF pulse amplitude

    b. Double the RF pulse width

---

* Some of the following questions are courtesy of Professor E. McVeigh. With permission.

   c. Cut the RF pulse amplitude in half

   d. Cut the RF pulse width by one-half

   e. a or b

   f. c or d

   g. None of the above

4. If the amplitude of the readout gradient is doubled (and all other parameters are held constant), the field of view of the resulting image in the readout direction is

   a. Doubled

   b. Cut in half

   c. Unchanged if the sampling rate is also cut in half

   d. Unchanged if the sampling rate is also doubled

   e. b and c

   f. a and d

5. Suppose a water sample is immersed in a static magnetic field B_o applied in the z direction. The time constant T_1:

   a. Characterizes the rate at which the magnetization vector tips away from the z axis under the influence of an applied RF pulse

   b. Is greater than T_2

   c. Characterizes the rate at which the longitudinal component of the magnetization M_z returns to its equilibrium value

   d. Both a and b

   e. Both b and c

6. Nonrecoverable loss of the existing transverse magnetization M_{xy} occurs due to

   a. Nonhomogeneous static magnetic fields

   b. Differences in the absorption of radio frequency energy at different parts of the sample

   c. T_1 decay

   d. T_2 decay

   e. Nonlinearities in the applied gradients

7. A slice is selected with a combined RF pulse and gradient pulse. In order to select a slice of the same thickness that is offset from the center of the imaging volume you need to

   a. Change the magnitude of the gradient

   b. Increase the power of the RF pulse

   c. Change the frequency of the RF pulse

   d. Both a and c

   e. Both b and c

8. A 10 mm slice from the center of a 10 cm diameter cylinder of water is selected with a $\pi/2$ pulse at 1.5 T. The free induction decay is sampled with 256 points in the presence of a 1 Gauss/cm gradient. If the points of the projection of the slice are to be separated by 0.5 mm, the separation of the sampled points of the free induction decay is

   a. 9.14 ms

   b. 4.70 ms

   c. 2.35 ms

   d. 18.4 µs

   e. 36.8 µs

9. If we cut the amplitude of the gradient in half, keeping all else the same, the separation of the points of the projection will now be

   a. 0.125 mm

   b. 0.25 mm

   c. 0.5 mm

   d. 1.0 mm

   e. 2.0 mm

10. Slice selection is achieved in an MRI experiment using a sinc RF pulse and a slice selection gradient $G_z = 1$ Gauss/cm. If the selection plane is desired to be shifted by +5 cm, you have to

    a. Increase G_z from 1 to 2 Gauss/cm

    b. Shift the RF excitation frequency by 21.3 KHz

    c. Shift the RF excitation frequency by 2.13 KHz

    d. Shift the RF excitation frequency by 4.26 KHz

11. The elicited and detected NMR signal from biological tissue ranges approximately within a range of a few:

    a. nV

    b. V

    c. μV

    d. mV

    e. Hundreds of volts

12. You want to tip the magnetization vector with the use of an RF pulse. After initial experiments on a phantom, it is observed that a rectangular pulse of a time duration of 1 ms and maximum amplitude A_{rf} *does not elicit an NMR signal* (no free induction decay) and that (on a separate experimental test) a second rectangular pulse of a duration of 0.5 ms with amplitude A_{rf} does generate an FID. It can be concluded that

    a. An RF pulse with a duration of 0.25 ms will generate an FID with an amplitude of $A_{rf}/2$

    b. An RF pulse with a duration of 2 ms and an amplitude of A_{rf} will not generate an FID

    c. A pulse with a duration of 1 ms and an amplitude of 2 A_{rf} will not generate an FID

    d. All the above

    e. b and c

13. Some of the advantages of MRI in comparison to other medical image diagnostic modalities include that it:

    a. Generates high-resolution images

    b. Can selectively excite spins in any plane orientation in three-dimensional space

    c. Exhibits excellent soft tissue contrast

    d. Has an associated low exam cost

    e. a, b, and c

    f. All of the above

14. Suppose that an aqueous solution in a plastic container is placed inside an MR scanner with a static field strength B_o for imaging. The T_1 relaxation for such a solution:

    a. Is characteristic of the rate at which the magnetization vector rotates around the Z axis toward the transverse plane XY, during the application of an excitation RF pulse

b. Is equal to or larger in value than the value of the transverse relaxation constant T_2

c. Is characteristic of the rate at which the magnetization vector component along the Z axis, M_z, is restored in its equilibrium value

d. a and b

e. b and c

15. Magnetic resonance was discovered by

a. Richard Ernst

b. Paul Lauterbur

c. Felix Bloch

d. Edward Purcell

e. a and c

f. c and d

16. Fat is present in most biological tissue (in the form of polymethyl (CH_3) organic compounds). If the resonance frequency of fat (methyl protons) differs by 200 Hz in comparison to the resonant frequency of tissue protons, in the resulting MR image, a chemical shift artifact arises. Such an artifact is due to

a. The T_2 difference of water and fat protons

b. Fat–water T_1 differences

c. Fat–water resonant frequency differences

d. Selection of the tip angle value

e. c and d

17. You are asked to conduct an *inversion recovery* experiment to measure a sample's T_1 value. The elicited $T_1 = 10$ ms. What is the inversion time (TI) for which *no NMR signal is elicited*?

a. 10.2 ms

b. 6.9 ms

c. 35 ms

d. 65 ms

e. 7.2 ms

18. In an MR experiment a slice-selective gradient pulse is employed to achieve slice selection. In order to achieve slice selection of another slice having the same slice thickness, spatially shifted from the magnet's iso-center, the following need to be changed:

   a. The slice selection gradient

   b. The radio frequency power increased

   c. The RF pulse

   d. a and b

   e. b and c

   f. a and b and c

Section II

Select true or false.

1. An MRI system uses:

   T/F   A static magnet that generates a magnetic field

   T/F   A rotating system of a camera with detectors

   T/F   Gradients in the three physical axes

   T/F   A radio frequency coil for signal detection

2. The explanation and understanding of the NRM phenomenon for proton nuclei based on the quantum mechanical theory considers:

   T/F   The presence of two energy levels of different energies

   T/F   The alignment of the magnetization vector of each nucleus antiparallel to the direction of the external magnetic field

   T/F   The alteration of the spin population distribution at the two energy levels in accordance to elicited thermal changes

   T/F   The theoretical solution of Schrödinger's equation for the estimation of the Hamiltonian function

3. To generate a T_1-weighted image:

   T/F   TR $\gg T_1$ of existing tissues within the field of view

   T/F   TR $\sim T_1$ of existing tissues within the field of view

   T/F   TE $\ll T_2$ existing tissues within the field of view

   T/F   TE $\sim T_2$ existing tissues within the field of view

4. The relaxation constant T_2

    T/F   Is known as the spin–lattice relaxation

    T/F   Has a value affected by local field inhomogeneities

    T/F   Has a value $\geq T_1$ for aqueous solution

    T/F   Determines the increase and restoration of the longitudinal magnetization vector to equilibrium

5. In regards to the NMR phenomenon:

    T/F   Higher field strengths yield higher SNR

    T/F   Occurrence of phase encoding pulses in spin–warp imaging can be executed during the same time period during which slice selection pulses are executed

    T/F   Fast gradient switching leads to eddy currents

6. T_1 relaxation:

    T/F   Is known as the spin–lattice relaxation

    T/F   Has a value affected by local static field inhomogeneities

    T/F   Has a value $\geq T_2$ in aqueous solutions

    T/F   Determines the rate and regrowth of the longitudinal magnetization vector to its equilibrium value

Solutions to Selected Problems

Chapter 1

Fourier Transformations

1. $g(x, y) = (x + \alpha y)^2$ (SQ 1.1)

    a. $g(x, y).\delta(x-1, y-2) = g(1,2) = (1 + \alpha.2)^2 = 1 + 4\alpha + 4\alpha^2$ (SQ 1.2)

    b. $g(x,y) * \delta(x-1, y-2) = \int_{-\infty}^{+\infty}\int_{-\infty}^{+\infty} g(x-\xi, y-\eta).\delta(\xi-1, \eta-2)d\xi \, d\eta$ (SQ 1.3)

       which reduces to $g(x-1, y-2) = [x-1 + \alpha(y-2)]^2$ (SQ 1.4)

2.

    a. $f(x,y) = f_1(x).f_2(y)$ (SQ 2.1)

       $g(x,y) = g_1(x).g_2(y)$ (SQ 2.2)

       $f(x,y) * * g(x,y) = \int_{-\infty}^{\infty}\int f(x-\kappa, y-\lambda)g(\kappa,\lambda)d\kappa \, d\lambda$ (SQ 2.3)

$$= \int_{-\infty}^{\infty}\int f_1(x-\kappa)f_2(y-\lambda)g_1(\kappa)g_2(\lambda)d\kappa \, d\lambda \qquad \text{(SQ 2.4)}$$

$$= \int_{-\infty}^{\infty}\int \left[f_1(x-\kappa)g_1(\kappa)d\kappa\right]\left[f_2(y-\lambda)g_2(\lambda)d\lambda\right] \qquad \text{(SQ 2.5)}$$

$$= \int_{-\infty}^{\infty} f_1(x-\kappa)g_1(\kappa)d\kappa \cdot \int_{-\infty}^{\infty} f_2(y-\lambda)g_2(\lambda)d\lambda \qquad \text{(SQ 2.6)}$$

$$= \varphi_1(x) \cdot \varphi_2(y)$$

$$= [f_1(x) * g_1(x)].[f_2(x) * g_2(y)] \qquad \text{(SQ 2.7)}$$

3. a. $F(\omega) = FT\{f(t)\} = \displaystyle\int_{-\infty}^{+\infty} f(t)e^{-j\omega t}\, dt$ 　　　　　　　(SQ 3.1)

$\quad = \displaystyle\int_{-T/2}^{T/2} Ae^{j\Omega t} - j\omega t\, dt = \dfrac{A}{(-j\omega)} e^{-j\omega T/2} - e^{j\omega T/2}$ 　　　(SQ 3.2)

$\quad = \mathrm{Re}\left\{ \dfrac{A}{\omega}\left(\dfrac{e^{j\omega T/2} - e^{-j\omega T/2}}{jT/2} \right)\dfrac{T}{2} \right\}$ 　　　　　(SQ 3.3)

$\quad = AT \sin c\left(\dfrac{\omega T}{2} \right)$ 　　　　　　　　　　　　(SQ 3.4)

See Figure Q3.1.

b.

$$f(x,y) = \begin{cases} A & -T/2 \le x, y \le T/2 \\ \\ 0 & \text{otherwise} \end{cases}$$　　　(SQ 3.5)

$$F(\omega_x, \omega_y) = FT\{f(x,y)\} = \int_{-\infty}^{+\infty}\int_{-\infty}^{+\infty} f(x,y)e^{-j(\omega_x x + \omega_y y)}\, dxdy$$　　(SQ 3.6)

$$= \int_{-\infty}^{+\infty} Ae^{-j\omega_x x}\, dx \int_{-\infty}^{+\infty} Ae^{-j\omega_y y}\, dy$$　　　(SQ 3.7)

$$= A\int_{-T/2}^{+T/2} e^{-j\omega_x x}\, dx \int_{-T/2}^{+T/2} e^{-j\omega_y y}\, dy$$　　　(SQ 3.8)

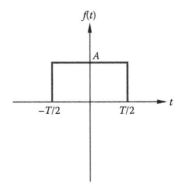

Figure Q3.1. Rectangular function in one dimension.

$$= \frac{AT}{\left(\frac{\omega_x T}{2}\right)} \sin\left(\frac{\omega_x T}{2}\right) \cdot \frac{T}{\left(\frac{\omega_y T}{2}\right)} \sin\left(\frac{\omega_y T}{2}\right) \qquad \text{(SQ 3.9)}$$

$$= AT^2 \sin c\left(\frac{\omega_x T}{2}\right) \sin\left(\frac{\omega_y T}{2}\right) \qquad \text{(SQ 3.10)}$$

See Figure Q3.2.

In a similar fashion, repeat for the sinc and the Gaussian functions.

4. b.

$$FT[I(\alpha x, \beta y)] = \int\int_{-\infty}^{\infty} I(\alpha x, \beta y) e^{-j(\omega_x x + \omega_y y)} dxdy \qquad \text{(SQ 4.1)}$$

$$Let\ \alpha x = X\ and\ \beta y = Y \qquad \text{(SQ 4.2)}$$

$$\Rightarrow dX = \alpha dx\ and\ dY = \beta dy \qquad \text{(SQ 4.3)}$$

$$FT[I(\alpha x, \beta y)] = \frac{1}{|\alpha\beta|} \int\int_{-\infty}^{\infty} I(X,Y) e^{-j\left(\frac{\omega_x}{\alpha} X + \frac{\omega_y}{\beta} Y\right)} dXdY = \frac{1}{|\alpha\beta|} \cdot I\left(\frac{\omega_x}{\alpha}, \frac{\omega_y}{\beta}\right) \qquad \text{(SQ 4.4)}$$

c. The image will be smoothed/blurred.

d. Please refer to example in Chapter 1, p. 11.

5. a. $g(m,n) = \dfrac{2}{243}$ \hfill (SQ 5.1)

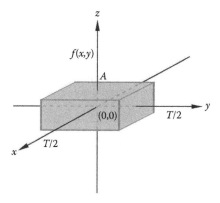

Figure Q3.2. Rectangular function in two dimensions.

b. $g(m-3,n-1) = \dfrac{(m-3) - \dfrac{2}{(n-1)}}{([m-3]-3[n-1])^2}$ (SQ 5.2)

c. $x[m, n] = 3\delta[m, n] + \delta[m - 1, n] + 2\delta[m - 1, n - 1]$ (SQ 5.3)

6. a. $-\dfrac{1}{27}$ (SQ 6.1)

b. $g(x,y) = \dfrac{1}{([x-3]-3[y-1])^3}$ (SQ 6.2)

Chapters 2–7 and 9: Magnetic Resonance

1. Use the equation

$$M_z(t) = M_o(1 - e^{-t/T_1}) + M_o \cos\alpha\, e^{-t/T_1}$$ (SQ 1.1)

The last term of the above equation denotes the initial magnetization after an α pulse. Let

$$M_z^n(t) = M_o(1 - e^{-t/T_1}) + M_z^{ss} \cos\alpha\, e^{-t/T_1}$$ (SQ 1.2)

where M_z^{ss} is the steady-state magnetization defined by

$$M_z^n(\mathrm{TR}) = M_z^{ss}$$ (SQ 1.3)

From (SQ 1.2) and (SQ 1.3) listed above,

$$M_z^{ss} = M_o \frac{(1 - e^{-\mathrm{TR}/T_1})}{(1 - \cos\alpha\, e^{-\mathrm{TR}/T_1})}$$ (SQ 1.4)

2. Let's denote the two tissues as A and B, respectively.

$$M_a(t) = M_a(1 - e^{-\mathrm{TR}/T_1})\, M_b(t) = M_b(1 - e^{-\mathrm{TR}/T_1})$$ (SQ 2.1)

where M_a, M_b are considered to be constant for a given MRI experiment using a given pulse sequence. Assuming that $M_a = M_b$ and that $T_1^b > T_1^a$ yields:

a. The expected signal difference is $\Delta S = M_a(\mathrm{TR}) - M_b(\mathrm{TR}) = M_o\left(e^{-\mathrm{TR}/T_1^b} - e^{-\mathrm{TR}/T_1^a}\right)$. For tissues A and B with T_1^A, T_1^B, to maximize the signal difference:

$$\frac{\partial(\Delta S)}{\partial \mathrm{TR}} = 0$$ (SQ 2.2)

or

$$\frac{\partial}{\partial \text{TR}}\left[e^{-\text{TR}/T_1^B} - e^{-\text{TR}/T_1^A}\right] = -\frac{1}{T_1^B}e^{-\text{TR}/T_1^B} + \frac{1}{T_1^B}e^{-\text{TR}/T_1^A} \qquad \text{(SQ 2.3)}$$

$$\therefore \frac{T_1^B}{T_1^A} = e^{-\text{TR}\left(\frac{1}{T_1^B} - \frac{1}{T_1^B}\right)} \qquad \text{(SQ 2.4)}$$

or

$$\text{TR} = \left(\frac{T_1^A T_1^B}{T_1^B - T_1^A}\right)\ln\left(\frac{T_1^B}{T_1^A}\right) \qquad \text{(SQ 2.5)}$$

We need to calculate

$$\frac{d(\Delta S)}{d\text{TR}} = 0 \Rightarrow \text{TR} = \left(\frac{T_1^a T_1^b}{T_1^b - T_1^a}\right)\ln\left(\frac{T_1^b}{T_1^a}\right) \qquad \text{(SQ 2.6)}$$

b. Recall that

$$N \equiv NEX = \textit{no. of averages} = \frac{\textit{Total imaging time}}{\text{TR}} = \frac{t_{tot}}{\text{TR}} \qquad \text{(SQ 2.7)}$$

$$\text{SNR} \propto \Delta S.\sqrt{NEX} = \frac{\Delta S\sqrt{t_{tot}}}{\sqrt{\text{TR}}} \qquad \text{(SQ 2.8)}$$

Maximize SNR if

$$\frac{\partial}{\partial \text{TR}}\left(\frac{\Delta S}{\sqrt{\text{TR}}}\right) = 0 \qquad \text{(SQ 2.9)}$$

or

$$\frac{\partial}{\partial \text{TR}}\left[\frac{e^{-\text{TR}/T_1^B} - e^{-\text{TR}/T_1^A}}{\sqrt{\text{TR}}}\right] = 0 \qquad \text{(SQ 2.10)}$$

which becomes

$$\left[-\frac{1}{T_1^B}e^{-\text{TR}/T_1^B} + \frac{1}{T_1^A}e^{-\text{TR}/T_1^A}\right]\sqrt{\text{TR}} - \frac{1}{2}\left[\frac{e^{-\text{TR}/T_1^B} - e^{-\text{TR}/T_1^A}}{\sqrt{\text{TR}}}\right] = 0 \Rightarrow \quad \text{(SQ 2.11)}$$

$$2\text{TR}\left[\frac{1}{T_1^A}e^{-\text{TR}/T_1^A} - \frac{1}{T_1^B}e^{-\text{TR}/T_1^B}\right] = e^{-\text{TR}/T_1^B} - e^{-\text{TR}/T_1^A} \qquad \text{(SQ 2.12)}$$

The answer to the transcendental equation SQ 2.12 follows using either a graphical or a numerical solution.

3. Use the following:

$$f(t) \Leftrightarrow F(t) \tag{SQ 3.1}$$

$$f\left(\frac{t}{a}\right) \Leftrightarrow |a| \, F(af) \tag{SQ 3.2}$$

$$e^{-\pi t^2} \Leftrightarrow e^{-\pi f^2} \tag{SQ 3.3}$$

$$F[f(2t)] \Leftrightarrow \frac{1}{2} F\left(\frac{f}{2}\right) \tag{SQ 3.4}$$

Hence,

$$A(t) \Leftrightarrow A_o e^{-(t^2/\sigma^2)} = A_o e^{-\pi\left(\frac{t}{\sqrt{\pi}\sigma}\right)^2} \Rightarrow \tag{SQ 3.5}$$

$$A(f) = A_o e^{-\pi^2 \sigma^2 f^2} \tag{SQ 3.6}$$

a. $e^{-\pi^2 \sigma^2 f_{1/2}^2} = \frac{1}{2} \Leftrightarrow f_{1/2} = \sqrt{\dfrac{\ln 2}{\pi^2 \sigma^2}}$ (SQ 3.7)

See Figure Q3.1.

But FWHM = $2f_{1/2}$, so FWHM ≈ 530 Hz. Also, $\Delta f = \dfrac{\gamma}{2\pi} G_z \Delta_z$ with Δf = 530 Hz. Additionally, G_z = 1G/cm, $\gamma/2\pi$ = 4.258 kHz/G.

$$\Rightarrow ST = \frac{\Delta f}{G_z\left(\dfrac{\gamma}{2\pi}\right)} = \frac{530}{1 \times \dfrac{10^{-4}}{10^{-2}} \times 42.58 \times 10^6} = 1.25\,\text{mm} \tag{SQ 3.8}$$

b. Halving G_z doubles Δ_z, i.e., ST_{new} = 0.25 cm.

c. Halving σ doubles $f_{1/2}$, so slice thickness doubles.

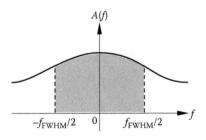

Figure Q3.1. Determination of FWHM from the frequency response of a Gaussian function.

d. When σ is halved,

$$A'(t) = A_o e^{-\left\{\frac{t^2}{\left(\frac{\sigma}{2}\right)^2}\right\}} = A_o e^{\left(-\frac{2t}{\sigma}\right)^2} \overset{FT}{\Longleftrightarrow} \frac{A_o}{2} e^{-\pi^2\left(\frac{f}{2}\right)^2 \sigma^2} \qquad \text{(SQ 3.9)}$$

The amplitude of the Gaussian profile halves in the frequency domain, so the bandwidth doubles.

e. The phase can be evaluated from

$$\omega(z) = -\gamma(B_o + G_z z) \qquad \text{(SQ 3.10)}$$

$$\varphi(z) = \omega(z)\Delta t_z = -\gamma(B_o + G_z z)\Delta t_z \qquad \text{(SQ 3.11)}$$

In most cases:

$$\varphi(z) = \int_0^T \gamma G_z(t) z \, dt \qquad \text{(SQ 3.12)}$$

Given a constant z gradient, then

$$\varphi(z) = kz \qquad \text{(SQ 3.13)}$$

where k is a constant.

4. Consider $s(n,m) = e^{-(nT+t_y)/T_2} \iint \rho(x,y) e^{-i\gamma(nTG_x x + mt_y \Delta_y y)} dx\, dy$ (SQ 4.1)

a. This term is representative of the decay of the transverse magnetization. Decay continues during A/D sampling until someone reaches k-space regions where only noise exists. The effect of this term e^{-t_y/T_2} (which has a constant value for a given $T_2 < 1$) is an equal attenuation throughout the image.

b. The presence of the term e^{-nT/T_2} emulates a low-pass filter in the x direction. Its effect is a convolution of a Lorentzian with the image domain, leading to blurring.

5. Treat (a), (b), (c), (d), and (e) in a similar fashion by using

$$\varphi_x(t) = \int_0^t \gamma G_x(t) x(t) \, dt \qquad \text{(SQ 5.1)}$$

General approach:

• To null the position term, we need two gradient lobes (one up and one down)

• To null the position and flow terms, use three lobes as in (b)

- To null the position, flow, and acceleration terms, we need one more lobe only.

d. Switch gradient on for the duration T for each setting of G_1, G_2, G_3, G_4 (Figure Q5.1).

$$\varphi(t) = \gamma \int_0^{4T} G_x(t)x(t)\,dt = \gamma \left[\begin{array}{l} \int_0^{T} G_1(x_0 + vt + \frac{1}{2}at^2)\,dt + \int_T^{2T} G_2(x_0 + vt + \frac{1}{2}at^2)\,dt \\[2ex] + \int_{2T}^{3T} G_3(x_0 + vt + \frac{1}{2}at^2)\,dt + \int_{3TA}^{4T} G_4(x_0 + vt + \frac{1}{2}at^2)\,dt \end{array} \right]$$

$$= x_0T + vT^2 + aT^3 \qquad\qquad (SQ\ 5.2)$$

$$= \text{position term} + \text{flow term} + \text{acceleration term}$$

e. We need to evaluate G_1, G_2, G_3, and G_4 to null all terms. Note that the solution is not unique. Repeat part (d) with three lobes of gradient pulse.

6. a. $FOV = \dfrac{1}{\left(\dfrac{\gamma}{2\pi}\right)GT} = 6.01\,\text{cm}$

with

$$\left(\frac{\gamma}{2\pi}\right) = 4.258\,\text{kHz/G}$$

$$\text{Sampling frequency} = \frac{N}{NT} = \frac{256}{10\,\text{ms}} = 25.6\,\text{kHz}$$

Bandwidth $\Delta f = (4.258\,\text{kHz/G}).(5\,\text{cm}).(1\,G/\text{cm}) = 21.3\,\text{kHz}$

b. Number of pixels $= \dfrac{5\,\text{cm}}{6.01\,\text{cm}} \cdot 256 = 213\,\text{pixels}$

c. $\Delta f' = (4.258\,\text{kHz/G})(5\,\text{cm})(0.5\,G/\text{cm}) = 10.6\,\text{kHz}$
   The FOV and the pixel number remain unchanged.

d. There is no change if no extra reconstruction technique is applied.

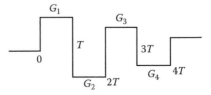

Figure Q5.1. Acceleration compensation sequence.

7. Start from

$$s(n,m) = e^{-(nT+t_y)/T_2} \iint \rho(x,y)e^{-i\gamma(nTG_xx+mt_y\Delta_yy)}dxdy \qquad \text{(SQ 7.1)}$$

Note that $\omega = 2\pi f = \gamma G_x$. So, $dx = \dfrac{2\pi}{\gamma G_x}df$. This leads to

$$s(t) = e^{-(t+t_d)/T_2} \int \int \left[\rho(x,y)e^{-i\gamma G_x xt}dxdy \right] e^{-i2\pi ft} \left(\frac{2\pi}{\gamma G_x} \right) df \qquad \text{(SQ 7.2)}$$

Taking the Fourier transform (or equivalently the inverse FT):

$$F[s(t)] = F\left[e^{-t/T2} \right] * \left[\int \rho(x,y)dy \right] \qquad \text{(SQ 7.3)}$$

Hence the Fourier (or inverse Fourier) transform of the signal is the convolution of a Lorentzian with the profile.

8. a. Phase encode in two dimensions and frequency encode in the third.

   b. The signal equation is an extension of Equation 4.25 in Chapter 4:

$$s(m,n,t) = e^{-t/T_2} \iiint \rho(x,y)\,dye^{-i\gamma(G_xxT+m\Delta G_y\Delta t_yy+n\Delta G_z\Delta t_zz)}dxdydz \qquad \text{(SQ 8.1)}$$

   c. One possible way is to use 3D Fourier reconstruction.

9. a. If M represents a magnetization vector, then M_n^b represents the longitudinal magnetization immediately before the nth RF pulse. Similarly, M_n^a represents the longitudinal magnetization immediately after the nth RF pulse. Immediately after the first RF pulse, the remaining longitudinal magnetization is

$$M_1^a = M_1^b \cos a = M_o \cos a \qquad \text{(SQ 9.1)}$$

   If one waits for a time TR before the next RF pulse, the longitudinal magnetization before the second pulse is calculated from the Bloch equations as

$$M_2^b = M_o \left(1 - e^{-\frac{TR}{T_1}} \right) + M_1^a e^{-\frac{TR}{T_1}} \qquad \text{(SQ 9.2)}$$

   Substituting SQ 9.1 into SQ 9.2 gives

$$M_2^b = M_o \left(1 - e^{-\frac{TR}{T_1}} \right) + M_1^b \cos a.e^{-\frac{TR}{T_1}} \qquad \text{(SQ 9.3)}$$

After the third RF pulse, the longitudinal magnetization becomes

$$M_3^b = M_o \left(1 - e^{-\frac{TR}{T_1}} \right) + M_2^b \cos a . e^{-\frac{TR}{T_1}} \qquad \text{(SQ 9.4)}$$

where M_2^b is given above in (SQ 9.2). Generalizing (SQ 9.4) for the $(n+1)$th pulse,

$$M_{n+1}^b = M_o \left(1 - e^{-\frac{TR}{T_1}} \right) + M_n^b \cos a . e^{-\frac{TR}{T_1}} \qquad \text{(SQ 9.5)}$$

If we assume that $M_{n+1}^b = M_n^b$ when n is large, then

$$M_{n+1}^b = \frac{M_o \left(1 - e^{-\frac{TR}{T_1}} \right)}{1 - \cos a . e^{-\frac{TR}{T_1}}} \qquad \text{(SQ 9.6)}$$

b. And the transverse magnetization immediately after the $(n+1)$th pulse is

$$M_{n+1}^t = \frac{M_o \left(1 - e^{-\frac{TR}{T_1}} \right) \sin a}{1 - \cos a . e^{-\frac{TR}{T_1}}} \qquad \text{(SQ 9.7)}$$

Equation SQ 9.7 can thus be rewritten as

$$M_{n+1}^b = M_o \left(1 - e^{-\frac{TR}{T_1}} \right) \left(1 + \cos a . e^{-\frac{TR}{T_1}} + \cos^2 a . e^{-\frac{2TR}{T_1}} + \dots + \left[\cos a . e^{-\frac{TR}{T_1}} \right]^{n-1} \right)$$

$$+ M_o \left(1 - e^{-\frac{TR}{T_1}} \right)^n \qquad \text{(SQ 9.8)}$$

or alternatively:

$$M_{n+1}^b = M_o \left(1 - e^{-\frac{TR}{T_1}} \right)^n + M_o \left(1 - e^{-\frac{TR}{T_1}} \right) \sum_{i=0}^{n-1} \left(\cos a . e^{-\frac{TR}{T_1}} \right)^i \qquad \text{(SQ 9.9)}$$

Using the geometric series equation,

$$\sum_{k=N_1}^{N_2} a^k = \frac{a^{N_1} - a^{N_2+1}}{1-a}, \quad N_2 \geq N_1, \ a < 1 \qquad \text{(SQ 9.10)}$$

Equation SQ 9.9 becomes

$$M_{n+1}^b = M_o\left(1 - e^{-\frac{TR}{T_1}}\right)^n + M_o\left(1 - e^{-\frac{TR}{T_1}}\right)\frac{\left(1 - \cos a.e^{-\frac{TR}{T_1}}\right)^n}{1 - \cos a.e^{-\frac{TR}{T_1}}} \quad \text{(SQ 9.11)}$$

describing the time evolution of the longitudinal magnetization with increasing repetitions. The transverse magnetization immediately after the $(n+1)$th RF pulse is $M_{n+1}^b \sin a$. Since

$$\left(1 - \cos a.e^{-\frac{TR}{T_1}}\right)^n = 0, \quad as\, n \to \infty \quad \text{(SQ 9.12)}$$

the longitudinal magnetization eventually reaches a steady state,

$$M^b = \frac{M_o\left(1 - \cos a.e^{-\frac{TR}{T_1}}\right)}{1 - \cos a.e^{-\frac{TR}{T_1}}} \quad \text{(SQ 9.13)}$$

as before.

c. To compute the optimum tip angle a_{opt}, take the first derivative of the solution from the first part of the problem and set it to zero,

$$\frac{\delta S}{\delta a} = M_o\left(1 - e^{-\frac{TR}{T_1}}\right)e^{-\frac{TE}{T_2}}\frac{\delta}{\delta a}\left(\frac{\sin a_{opt}}{1 - \cos a.e^{-\frac{TR}{T_1}}}\right) = 0 \quad \text{(SQ 9.14)}$$

and

$$\frac{\delta}{\delta a}\left(\frac{\sin a_{opt}}{1 - \cos a.e^{-\frac{TR}{T_1}}}\right) = \frac{\left(\cos a_{opt} - \cos^2 a_{opt}e^{-\frac{TR}{T_1}}\right) - \left(e^{-\frac{TR}{T_1}}\sin^2 a_{opt}\right)}{\left(1 - \cos a.e^{-\frac{TR}{T_1}}\right)^2} = 0 \quad \text{(SQ 9.15)}$$

This means that

$$\left(\cos a_{opt} - \cos^2 a_{opt}e^{-\frac{TR}{T_1}}\right) - \left(e^{-\frac{TR}{T_1}}\sin^2 a_{opt}\right) = 0 \quad \text{(SQ 9.16)}$$

or equivalently:

$$\cos a_{opt} - e^{-\frac{TR}{T_1}}\left(\cos^2 a_{opt} + \sin^2 a_{opt}\right) = \cos a_{opt} - e^{-\frac{TR}{T_1}} = 0 \quad \text{(SQ 9.17)}$$

Therefore,

$$a_{opt} = \cos^{-1} e^{-\frac{TR}{T_1}} \qquad \text{(SQ 9.18)}$$

An optimum tip angle is necessary because if a is close to zero, no magnetization is being tipped into the transverse plane. If $a=90°$, then all the magnetization is "used up" in the first pulse. A balance between T_1 recovery and decrementing the longitudinal magnetization by RF pulses can be achieved, and is optimal at the Ernst angle.

d. When averaging is involved, $\text{SNR} \propto \sqrt{N_a}$, where N_a is the number of averages. Since the total scan time, T_s, remains constant, it equals $T_s = N_a TR$. Therefore, $\text{SNR} \propto \dfrac{1}{\sqrt{TR}}$, and to normalize for constant scan time, divide the derived expression for signal by $\sqrt{TR}$.

A graphical solution indicates clearly that the optimum TR approaches zero. This means that the pulse sequence should be run as fast as possible at the Ernst angle, and the maximum averaging should be used in the amount of the time available. It is impossible to reach the maximum, since the imaging gradients have slew rate and the maximum amplitude limitations, and the receiver bandwidth cannot be made infinitely high.

e. i.   Starting from Equation SQ 9.11,

$$M_{n+1}^b = M_o \left(1 - e^{-\frac{TR}{T_1}} \right)^n + M_o \left(1 - e^{-\frac{TR}{T_1}} \right) \left(\frac{1 - \cos a . e^{-\frac{TR}{T_1}}}{1 - \cos a . e^{-\frac{TR}{T_1}}} \right)^n \qquad \text{(SQ 9.19)}$$

describing the time evolution of the longitudinal magnetization with increasing repetitions. The transverse magnetization immediately after the $(n+1)$th RF pulse is $M_{n+1}^b \sin a$. Since

$$(\cos a . e^{-t/T_1}) = 0, \quad \text{as} \, n \to \infty \qquad \text{(SQ 9.20)}$$

the longitudinal magnetization eventually reaches a steady state,

$$M^b = \frac{M_o \left(1 - \cos a . e^{-\frac{TR}{T_1}} \right)}{1 - \cos a . e^{-\frac{TR}{T_1}}} \qquad \text{(SQ 9.21)}$$

as before.

ii. The total signal can be maximized by summoning the total transverse magnetization, and then determining the tip angle that maximizes that sum.

$$S = \left(\sum_{i=0}^{n-1} M_{i+1}^b \right) \sin a = \frac{M_o e^{-\frac{TR}{T_1}}(1-\cos a)}{1-\cos a.e^{-\frac{TR}{T_1}}} \sum_{i=0}^{n-1} \left(\cos a e^{-\frac{TR}{T_1}} \right)^i + \frac{nM_o \left(1 - e^{-\frac{TR}{T_1}} \right) \sin a}{\left(1 - \cos a.e^{-\frac{TR}{T_1}} \right)}$$

(SQ 9.22)

The first term can be written in closed form using the infinite series relationship, and then the first derivative of S is taken with respect to a. The algebra from here is long and very complicated and need not be solved.

10. a. $\rho(x, y) = \rho_{protons}(x, y) + \rho_{fat}(x, y) e^{i\Delta\omega nT}$ (SQ 10.1)

Hence:

$$s(n,m) = \iint \rho_{protons}(x,y) e^{-i\gamma(nTG_x x + mt_y \Delta G_y y)} dx dy$$ (SQ 10.2)

$$+ \iint \rho_{fat}(x,y) e^{-i\gamma nTG_x \left(x + \frac{\Delta\omega}{\gamma G_x} \right)}.e^{-i\gamma mt_y \Delta G_y y} dx dy$$ (SQ 10.3)

b. $s(n,m) = e^{-(nT + t_y)/T_2} \iint \rho(x,y) e^{-i\gamma(nTG_x x + mt_y \Delta G_y y)} dx dy$ (SQ 10.4)

c. Assume that

$$x = x_o + vt$$ (SQ 10.5)

where v is the velocity of the readout. Then

$$\rho(x, y) = \rho(x - (x_o + vt), y)$$ (SQ 10.6)

Thus the equation for a point object becomes

$$s(n,m) = e^{-i\gamma nTG_x(x_o + vt)}$$ (SQ 10.7)

In effect, the response is the multiplication of two decaying exponentials, with one of the two being dependent on velocity. Invoking the FT properties, multiplication in k-space is equivalent to convolution in the spatial (image) domain.

11. Excitation of two nonoverlapping slices in space can be achieved using a sinc RF pulse, modulated by a frequency f. The frequency f is the "beat" frequency as a result of the two modulation frequencies used to spatially select the two slices. See also the example in Chapter 4, p. 44.

12. a. Start from

$$s(t) = e^{-(t+t_d)/T_2} \iint \rho(x, y)\, dx\, dy \qquad \text{(SQ 12.1)}$$

If you are only dealing with a point object at the origin, then

$$\rho(x, y) = \delta(x, y) \qquad \text{(SQ 12.2)}$$

So,

$$s(t) = e^{-(t+t_d)/T_2} \qquad \text{(SQ 12.3)}$$

where t_d is the time delay between the center of the RF and the beginning of the FID. For the purpose of this exercise this effect is neglected, so:

$$s(t) = e^{-t/T_2} \qquad \text{(SQ 12.4)}$$

Taking the Fourier transform of $s(t)$,

$$S(\omega) = \int_{-\infty}^{\infty} s(t)e^{-j\omega t}\, dt = \int_{0}^{\infty} e^{-t/T_2} e^{-j\omega t}\, dt \qquad \text{(SQ 12.5)}$$

$$= -\frac{1}{\dfrac{1}{T_2} + j\omega}\left[e^{-\left(j\omega + \frac{1}{T_2}\right)t} \right]_{0}^{\infty} \qquad \text{(SQ 12.6)}$$

$$= \frac{1}{\dfrac{1}{T_2} + j\omega} = \frac{\dfrac{1}{T_2} - j\omega}{\left(\dfrac{1}{T_2}\right)^2 + \omega^2} \qquad \text{(SQ 12.7)}$$

b. i.   The real part of the frequency spectrum $S(\omega)$ is known as the absorption line shape and is given by

$$\mathrm{Re}\left[S(\omega)\right] = \frac{\dfrac{1}{T_2}}{\left(\dfrac{1}{T_2}\right)^2 + \omega^2} \qquad \text{(SQ 12.8)}$$

The imaginary part of the frequency spectrum $S(\omega)$ is known as the dispersion line shape and is given by

$$\mathrm{Im}\left[S(\omega)\right] = -\frac{\omega}{\left(\dfrac{1}{T_2}\right)^2 + \omega^2} \qquad \text{(SQ 12.9)}$$

ii. Power spectrum:

$$P(\omega) = \sqrt{\operatorname{Re}\left[S(\omega)\right]^2 + \operatorname{Im}\left[S(\omega)\right]^2} \qquad \text{(SQ 12.10)}$$

$$= \frac{1}{\sqrt{\left(\dfrac{1}{T_2}\right)^2 + \omega^2}} \qquad \text{(SQ 12.11)}$$

Thus,

$$\frac{1}{\left(\dfrac{1}{T_2}\right)^2 + \omega^2} = \frac{T_2^2}{4} \qquad \text{(SQ 12.12)}$$

So,

$$\omega_{HM} = \frac{\sqrt{3}}{T_2} \qquad \text{(SQ 12.13)}$$

and

$$f_{FMHM} = \frac{\sqrt{3}}{\pi T_2} \qquad \text{(SQ 12.14)}$$

c. i.   To find the FWHM, we need to find the maximum height of the absorption line width. For $\omega = 0$,

$$Max\|\operatorname{Re}(S(\omega)\| = T_2 \qquad \text{(SQ 12.15)}$$

Thus to find ω_{HM}, need to solve

$$\frac{\dfrac{1}{T_2}}{\left(\dfrac{1}{T_2}\right)^2 + \omega_{HM}^2} = \frac{T_2}{2} \qquad \text{(SQ 12.16)}$$

or

$$\frac{2}{T_2^2} = \frac{1}{T_2^2} + \omega_{HM}^2 \qquad \text{(SQ 12.17)}$$

Hence,

$$\omega_{HM}^2 = \frac{1}{T_2^2} \qquad \text{(SQ 12.18)}$$

Thus,

$$\omega_{HM} = \pm \frac{1}{T_2}$$ (SQ 12.19)

So it follows that

$$f_{FMHM} = \frac{1}{\pi T_2}$$ (SQ 12.20)

Note: The same answer follows from manipulation of the complex frequency spectrum.

Similarly, the FWHM for the power spectrum is found to be

$$f'_{FWHM} = \frac{\sqrt{3}}{\pi T_2}$$

ii. MR images are magnitude images; this imposes a limitation to the image spatial resolution.

If $T_2 = 50\,\text{ms}$, then $f_{FMHM} = 6.4\,\text{Hz}$ and $f'_{FMHM} = 11.1\,\text{Hz}$

d. The term e^{-t/T_2} acts as a low-pass filter in the x direction, and its effect is a convolution of the Lorentzian function with the image domain, leading to blurring.

13. a.

$$s(t) = \frac{1}{2\pi} \int_{-\omega_A}^{\omega_A} e^{j\omega t} d\omega$$ (SQ 13.1)

$$= \frac{\sin \omega_A t}{\pi t}$$ (SQ 13.2)

$$= \frac{\omega_A}{\pi} \sin c\left(\frac{\omega_A}{\pi}\right)$$ (SQ 13.3)

where $\omega_A = 4\pi \times 10^3$ rad/s.

b. Using the duality property of FT (Figure Q13.1):

c. Use a modulating frequency of $f = 4\,\text{kHz}$.

d. Truncation results in Gibb's ringing since the high-frequency components are eliminated. A way to alleviate such problems is to multiply the truncated sinc pulse $s'(t)$ with a windowing function (such as Hamming). In that way the slice profile is smooth and free of any ringing effects.

14. a. $\rho(x, y) = (x - A \sin \omega_o t, y)$

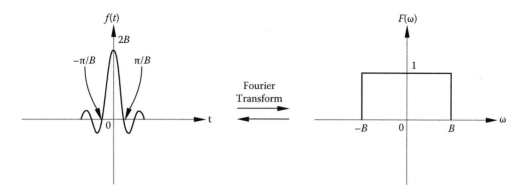

Figure Q13.1. Duality property of FT—the example of the sinc-rectangular function transformation.

Substitution in the imaging equation leads to

$$s(n,m) = e^{-(nTt+t_y)/T_2} e^{-i\gamma nTG_x A \sin\omega_o t} \qquad \text{(SQ 14.1)}$$

If the T_2 decay is neglected, then

$$s(n,m) = e^{-i\gamma nTG_x A \sin\omega_o t} \qquad \text{(SQ 14.2)}$$

b.

    i. Blurring in the readout

    ii. Artifacts in the phase encoding direction

18. a. Please refer to Sections 9.3 and 9.4.

    b. Oblique short axis.

    c. i. $SNR_{1,1} = \dfrac{150}{23}; SNR_{2,1} = \dfrac{23}{19}; SNR_{3,1} = \dfrac{15}{11}$

$$SNR_{1,2} = \frac{134}{15}; SNR_{2,2} = \frac{20}{18}; SNR_{3,2} = \frac{11}{14}$$

$$SNR_{1,3} = \frac{172}{40}; SNR_{2,3} = \frac{32}{17}; SNR_{3,3} = \frac{9}{8}$$

$$SNR_{1,4} = \frac{190}{54}; SNR_{2,4} = \frac{28}{33}; SNR_{3,4} = \frac{14}{16}$$

$$SNR_1 = \frac{SNR_{11} + SNR_{12} + SNR_{13} + SNR_{14}}{4}$$

      ii. $CNR_{myo-blood_1} = \dfrac{S_{myo} - S_{blood}}{\left(\dfrac{S_{myo} + S_{blood}}{2}\right)} = \dfrac{150 - 23}{\left(\dfrac{173}{2}\right)}$

19. a. $s(t) = \dfrac{M_o e^{-\text{TE}/T_2}(1 - e^{-\text{TR}/T_1})\sin\alpha}{(1 - e^{-\text{TR}/T_1})\cos\alpha}$ (SQ 19.1)

$= \dfrac{100 e^{-25/25}(1 - e^{-25/1000})\sin 30}{(1 - e^{-25/1000})\cos 30}$ (SQ 19.2)

$= \dfrac{100 e^{-1}(1 - e^{-1/40})\left(\dfrac{1}{2}\right)}{(1 - e^{-1/40})\left(\dfrac{\sqrt{3}}{2}\right)}$

b. Either the relaxation time T_1 or the relaxation time T_2.

c. i.   Assume that only T_2 changes.

$\dfrac{s'(t)}{s(t)} = 1.25 = \dfrac{e^{-\text{TE}/T_2'}}{e^{-\text{TE}/T_2}} \Rightarrow e^{-\text{TE}\left(\frac{1}{T_2'} - \frac{1}{T_2}\right)} = 1.25$ (SQ 19.3)

$\Rightarrow -\text{TE}\left(\dfrac{1}{T_2'} - \dfrac{1}{T_2}\right) = \ln(1.25)$

$\dfrac{1}{T_2} - \dfrac{1}{T_2'} = \dfrac{\ln(1.25)}{25 \times 10^{-3}}$

$\dfrac{1}{25 \times 10^{-3}}(1 - \ln[1.25]) = \dfrac{1}{T_2'}$

ii. Assume that only T_1 changes:

$s'(t) = 1.25 s(t)$

$\dfrac{s'(t)}{s(t)} = \dfrac{M_o e^{-\text{TE}/T_2}(1 - e^{-\text{TR}/T_1'})\sin\alpha(1 - e^{-\text{TR}/T_1}\cos\alpha)}{M_o e^{-\text{TE}/T_2}(1 - e^{-\text{TR}/T_1})\sin\alpha(1 - e^{-\text{TR}/T_1'}\cos\alpha)}$ (SQ 19.4)

$\Rightarrow \dfrac{1.25(1 - e^{-\text{TR}/T_1})}{\left(1 - e^{-\text{TR}/T_1}\dfrac{\sqrt{3}}{2}\right)} = \dfrac{(1 - e^{-\text{TR}/T_1'})}{(1 - e^{-\text{TR}/T_1'}\cos\alpha)}$ (SQ 19.5)

d. Shield, gradient, magnet, RF, RF coil, patient table, MRI suite, film—computer reconstruction, gradient amplifier, gradient pulse, computer, receiver, digitizer, pulse programmer, RF amplifier, RF source.

e. Please refer to Figure 7.1 Section 7.2.

f. i.   See Figure Q19.1.

ii. There is an excess 1 ppm of nuclei in the lower energy level under equilibrium conditions.

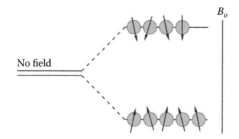

Figure Q19.1. Energy levels of the ^{1}H nucleus in the absence and presence of an external, static magnetic field.

iii. $$\dfrac{n_m}{n_{-m}} = e^{-\left(\frac{\Delta E}{kT}\right)}$$ (SQ 19.6)

    iv. 1. Conduct the experiment under lower temperatures.

        2. Increase the static magnetic field.

        3. Administer a contrast agent.

        4. Conduct a prepolarization experiment before imaging.

20. b. Computer tomography.

    c. Back-projection reconstruction based on the Fourier slice theorem.

Chapters 7 and 10: Spectroscopy and Instrumentation

3. The solution for computation of the rectangular coil (in a similar fashion to the cylindrical case) for B_z can be found in numerous electromagnetic or physics books.

    For a cylindrical coil of radius α, it can be proved that the field along the z direction B_z is

$$B_z = \frac{\mu_o I}{4\pi}\frac{8}{(z^2 + a^2)^{1/2}}$$ (SQ 3.1)

a. $a = radius = d/2 = 10\,cm$

$$B_z = \frac{\mu_o I a^2}{2(a^2 + z^2)^{3/2}}$$ (SQ 3.2)

$$P_{rms} = 2000\ W = I_{rms}^2 R = I_{max}^2 R \Rightarrow I_{max} = \sqrt{\frac{2P_{rms}}{R}} = 8.94\ A$$ (SQ 3.3)

$$B_{1,max}(z = 0) = \frac{\mu_o I_{max}}{2a} = 56.17\,\mu T$$ (SQ 3.4)

b. $B_{1,\text{max}}(z = 20\,cm) = \dfrac{\mu_o I_{\text{max}} a^2}{2(5a^2)^{3/2}} = 5.024\,\mu T$ (SQ 3.5)

c. $Power\,deposited = \dfrac{I_0^2 R_c}{I_0^2 (R_s + R_c)} = 1 - \left(\dfrac{Q_L}{Q_{unl}}\right) = 1 - \left(\dfrac{100}{200}\right) = 50\%$

(SQ 3.6)

Answers to Multiple Choice Questions

Section I

1. d
2. e
3. g
4. e
5. c
6. d
7. c
8. d
9. d
10. b
11. c
12. e
13. e
14. e
15. f
16. c

    17. b

    18. c

Section II

    1. T/F/T/T

    2. T/F/T/T

    3. F/T/F/F

    4. T/T/F/F

    5. T/F/T

    6. T/F/T/T

Glossary

Angiography: (Greek: *Angio-grafima.*) The representation of blood vessels using imaging.

Anterior: In front of; anatomically relevant descriptive term to denote tissue or organ position with respect to the reference coordinate system.

Artifact: A fictitious feature that appears on the reconstructed image that is not part of (or present in) the imaged anatomy. Often the end result of imperfect hardware, software, or due to physiological factors (breathing, cardiac motion) or the presence of external (to the human body) objects (e.g., implants, etc.).

Axial: (Greek: *Aksoniki*, "transaxial.") One of the three tomographic imaging right-angled planes that bisect the human body (together with sagittal and coronal). In this case, the resulting image depicts the human body into top and bottom (and right and left) parts.

Birdcage coil: A specific type of transmit-receive radio frequency coil with a cylindrical shape and parallel leg conducting elements, associated with increased B_1 field homogeneity. Its design resembles a constructed birdcage—hence its name.

Chemical shift: The inherent property of each nucleus to be associated with a resonance frequency depending on the nature of its local surrounding chemical environment. Chemical shift is often reported in parts per million (ppm), with the absolute zero reference being the resonant frequency of the chemical compound trimethylsilane (TMS).

Chemical shift imaging: The chemical imaging of a specific chemical moiety or nucleus, based on its spectral selection from a reconstructed spectrum using filters.

Coil: Term to describe the electrical resonant circuit that generates the weak magnetic field B_1. It can act as both a transmit coil and a receive coil. In its receive mode of operation (acts like an antenna) the coil receives the weak MR signal from tissue-emanating spins.

Continuous wave (CW): Original methodology with which spectroscopy was conducted (as compared to pulsed wave NMR), where a constant amplitude (excitation) electromagnetic wave is applied.

Contrast: (Greek: "Antithesis.") Mathematical definition expressing quantitatively the observer's ability to distinguish one tissue/organ from a neighboring one (or from the image background).

Contrast agent: A chemical substance (typically in the form of a solution) administered via the venous system into the organism. The substance changes the relaxation values T_1 and T_2 of the tissue of interest, thereby increasing the contrast of the reconstructed image.

Convolution: Mathematical integral operation between two functions.

Coronal: (Greek: *Stephaniaia,* "coronary," "on top.") Second of three right-angled tomographic imaging planes bisecting the body into head and foot and left and right parts.

Diffusion imaging: Imaging completed with a diffusion-sensitized gradient lobe (positioned between two 180° RF pulses) where the resulting image contrast is directly related to the diffusion coefficient of the water molecules.

Doubly balanced mixer: An electrical circuit device that modulates or demodulates (upconverts or downconverts) the detected signal (to the desired frequency or to baseband reference frequency).

Echo: A form of elicited magnetic resonance signal that stems from the refocusing of transverse magnetization. (Refer to Hahn's echoes.)

Echo planar imaging (EPI): (Following the invention by Sir Peter Mansfield.) An extremely fast MRI sequence that acquires a complete image in short times (of the order of a few milliseconds).

Echo time (TE)*:* The time interval between the 90° excitation pulse and the maximum amplitude of the echo signal in a spin–echo experiment.

Fast spin–echo: A multiple-echo, spin–echo acquisition with a train of 180° pulses following the initial 90° excitation RF pulse and a single readout gradient.

Field of view: The spatial extent of the image in both x and y directions (which may be different in the two directions). Typically reported in centimeters or millimeters.

Fourier transform (FT): A mathematical (integral) transformation (after the French mathematician Fourier) that allows conversion of a time domain function (or signal) into a frequency domain signal, and vice versa.

Fractional echo image: An imaging technique that takes advantage of k-space symmetry to allow image reconstruction using only a fraction of k-space data.

Free induction decay (FID): One form of elicited magnetic resonance imaging signal due to the decaying transverse magnetization.

Frequency encoding gradient (G_f): A gradient pulse that allows encoding of spin frequency with its spatial position (typically along the physical x axis).

Functional magnetic resonance imaging (fMRI): An MRI technique with which the elicited signal encodes functional activity of cortical centers, in accordance to the deoxyhemoglobin saturation of the arteriovenous system. (Refer to Ogawa et al. 1993.)

Gradient (G): A variation (or perturbation) of locally generated magnetic fields along the three spatial directions (x, y, and z). In MRI, such variations are typically linear in space.

Gradient echo: A form of an elicited MR signal due to the refocusing of transverse magnetization in response to the application of magnetic field gradient.

Gyromagnetic ratio: A nuclear constant value (associated with the symbol γ) that has a magnetic field dependence.

Imaging sequence: The collective sequence of RF and gradient pulses that allow generation of an MR image.

Inferior: Behind; anatomically relevant descriptive term to denote tissue or organ position with respect to the reference coordinate system.

Inversion recovery sequence: A specific type of MRI pulse sequence that uses a 180° inversion RF pulse to originally invert the magnetization vector, followed by the application of a 90° RF pulse to interrogate the magnitude of the magnetization at an interval TI, known as the inversion interval.

Inversion time (TI): (*See also* inversion recovery sequence.) The time interval between the inversion pulse and the sampling 90° RF pulse, in an inversion recovery sequence.

Isocenter: The location inside the bore of an imaging scanner (typically the [0, 0, 0] reference coordinate position) with a magnetic field strength B_o.

k-Space: A term introduced by Dr. Truman Brown, denoting the MR image space as represented by the time and phase raw data. The Fourier transform of k-space is the magnetic resonance image.

Larmor frequency: The frequency of precession of a nuclear spin upon exposure of the nucleus to an external magnetic field. The frequency is given by the product of the gyromagnetic ratio and the field strength of the external magnetic field.

Longitudinal magnetization: The component of the total magnetization along the B_o direction (z direction).

Lorentzian line shape: A mathematical function resulting from the Fourier transform of a decaying exponential signal.

Magnetic resonance angiography (MRA): (*See also* angiography.) An MRI technique used for angiographic imaging.

Magnetic resonance elastography (MRE): (Refer to work by Dr. Richard Ehman.) An MRI technique that uses concurrent ultrasonic wave excitation to allow shear modulus imaging of tissues/organs.

Magnetization transfer: A dedicated preimaging sequence of pulses that ensures transfer of net magnetization from one species to another in the cases where such species are associated with overlapping spectral positions.

Net magnetization vector: In the classical description of NMR, it represents the vectorial sum of all magnetization vectors associated with the ensemble of nuclear spins in a system.

Oblique imaging: Descriptive term to denote the imaging plane position along arbitrary orientations in three-dimensional space (other than the three right-angled planes).

Parallel imaging: An imaging technique that utilizes the B_1 field maps of receive-only phased array coils to allow reconstruction of an MR image.

Phantom: Typically a plastic or Plexiglas construction filled with aqueous salt solutions (often doped with nickel, iron, chromium, or other transitional metals) that can be imaged to test the performance of an MRI system, an RF coil, or a pulse sequence.

Phased array coil: Introduced by Dr. Peter Roemer in MRI, it is the specialized RF coil composed of at least two overlapping coil loops.

Pixel: The smallest image matrix element; also known as the picture element. (*See also* voxel.)

Posterior: Toward the back; anatomically relevant descriptive term to denote tissue or organ position with respect to the reference coordinate system.

Pulse sequence: The collective sequence of RF and gradient pulses that allow generation of an MR image. (*See also* imaging sequence.)

Quadrature detection: The simultaneous detection (using specialized phase shifting circuitry—similar to demodulation circuitry used in communication systems) of the M_x and M_y components of the magnetization vector with time.

Radio frequency: A frequency band in the electromagnetic spectrum with frequencies that range between a few tens and hundreds of millions of Hertz (MHz).

Raw data: The sampled values of the elicited transverse magnetization as a function of phase and time from an imaging sequence, also known as the *k*-space data. (*See also* term *k*-space.)

Relaxivities: The inverse values of the spin–lattice and spin–spin relaxation times.

Repetition time: The time between the centers of successive RF pulses in an imaging pulse sequence.

RF pulse: A shaped burst of RF energy used to excite the nuclear spins to a higher energy level.

Rotation matrix: A matrix used to describe the relation between the physical and gradient axes.

ρ-*Weighted image:* An MRI that is weighted to the proton density content of the imaged slice.

Saddle coil: A specific coil geometry/design with two conducting loops wrapped around opposite sides of a cylinder, spanning an arc/sector.

Sagittal: (Greek: *Oveliaia*, "sagittal.") One of the three tomographic imaging right-angled planes that bisect the human body (together with axial and coronal). In this case the resulting image depicts the human body in top and bottom (and head and foot) parts.

Sinc pulse: An RF pulse shaped in accordance to the mathematical function $\sin(x)/x$.

Slew rate: The rate at which a gradient may be turned on or off. The faster the slew rate, the shorter the minimum attainable TE value.

Solenoid coil: A transmit and receive RF imaging coil that has a generic cylindrical shape.

Specific absorption rate (SAR): Quantitative estimate of RF energy deposition (in watts) per kilogram of excited tissue in an MRI experiment.

Spin density: The function that describes the spatial distribution of spins in two- or three-dimensional space.

Spin–echo: (Also known as Hahn's spin–echo.) An MRI pulse sequence whose elicited echo signal arises from the refocusing of magnetization, from the application of a 90° and 180° RF pulses.

Spin–lattice relaxation: The process with which there is regrowth of the magnetization vector along the longitudinal axis.

Spin–lattice relaxation time (T_1): (Also known as the longitudinal magnetization T_1.) The physical quantity (time constant) that describes the exponential regrowth of the magnetization along the longitudinal axis (z axis).

Spin–spin relaxation: The process with which there is decay of the transverse magnetization.

Spin–spin relaxation time: The time constant characteristic of the exponential decay of the transverse magnetization with time.

Superconducting: The ability of certain materials (or mixtures thereof) to have an almost zero electrical resistance when they operate at temperatures close to absolute zero.

Superior: Toward the head; anatomically relevant descriptive term to denote tissue or organ position with respect to the reference coordinate system.

Surface coil: A receive-only RF coil that is typically placed on the surface of the imaged object.

T_1-Weighted image: An MR image where contrast is primarily dependent on the T_1 value(s) of the imaged tissue/organ.

T_2*: Pronounced as T-2-star, it is the modified spin–spin relaxation time as determined by the intrinsic T_2 value and local magnetic field inhomogeneities.

T_2-Weighted image: An MR image where contrast is primarily dependent on the T_2 value(s) of the imaged tissue/organ.

Thickness: Often referred to as the image slice or cross-sectional thickness.

Timing diagram: The pulse diagram representation of the imaging gradients and excitation radio frequency pulse with respect to time.

Tomographic: (Greek: *Tomy*, "cross-section," "slice.") Descriptive term to describe the anatomical slice to be imaged or the type of modality used for imaging.

Transverse magnetization: The component of the total magnetization vector (M_o) that is tipped (has been rotated) on the transverse (i.e., the *XY*) plane.

Variable bandwidth imaging: Imaging completed with different reception bandwidths (and hence different digitization rates) to allow control of the noise and chemical shift artifact.

Volume imaging: Three-dimensional representation of the body/organ structure. Typically achieved by phase encoding in two dimensions and signal readout in the third.

Voxel: The smallest cuboid element of the image matrix. Alternatively, it is also known as the volume element.

Bibliography

Abraham A. *The Principles of Nuclear Magnetism*. Oxford University Press, 1972, Oxford.

Allman T, Holland GA, Lenkinski RE, Charles HC. A Simple Method for Processing NMR Spectra in Which Acquisition Is Delayed: Applications to *In Vivo* Localized 31P NMR Spectra Acquired Using the DRESS Technique. *Magnetic Resonance in Medicine* 1988; 7:88–94.

Andrew ER, Bottomley PA, Hinshaw WS, Holland GN, Moore WS, Simaroj C. NMR Images by the Multiple Sensitive Point Method: Application to Larger Biological Systems. *Physics in Medicine and Biology* 1977; 22(5):971–974.

Barker, P. MRI Lecture Notes. Johns Hopkins University, 1992, Baltimore.

Berendsen HJC, Edzes HT. The Observation and General Interpretation of Sodium Magnetic Resonance in Biological Material. *Annals of the New York Academy of Sciences* 1973; 204:459–485.

Bernstein MA, King KF, Zhou XJ. *Handbook of MRI Pulse Sequences*. Elsevier, 2004, Amsterdam.

Blaimer M, Breuer F, Mueller M, Heidemann RM, Griswold MA, Jakob PM. SMASH, SENSE, PILS, GRAPPA: How to Choose the Optimal Method. *Topics in Magnetic Resonance Imaging* 2004; 15(4):223–236.

Blamire AM. The Technology of MRI—The Next 10 Years? *British Journal of Radiology* 2008; 81:601–617.

Bloch F. Nuclear Induction. *Physical Review* 1946; 70(7–8):460–479.

Bloch F, Hansen WW, Packard M. Nuclear Induction. *Physical Review* 1946; 69(3–4):127.

Bloembergen N, Purcell EM, Pound RV. Relaxation Effects in Nuclear Magnetic Resonance Absorption. *Physical Review* 1948; 73(7):679–712.

Boada FE, Gillen JS, Shen GX, Chang SY, Thulborn KR. Fast Three Dimensional Sodium Imaging. *Magnetic Resonance in Medicine* 1997; 37(5):706–715.

Bottomley PA. State of the Art. Human *In Vivo* NMR Spectroscopy in Diagnostic Medicine: Clinical Tool or Research Probe? *Radiology* 1989; 170:1–15.

Bottomley, PA. MR Spectroscopy of the Human Heart: The Status and the Challenges. *Radiology* 1994; 191(3):593–612.

Bottomley PA. Lecture Notes: Introduction to Magnetic Resonance in Medicine Course—Spectroscopy. Johns Hopkins University, 2003, Baltimore.

Bottomley PA, Foster TH, Argersinger RE, Pfeifer LM. A Review of Normal Tissue Hydrogen NMR Relaxation Times and Relaxation Mechanisms from 1–100 MHz: Dependence on Tissue Type, NMR Frequency, Temperature, Species, Excision and Age. *Medical Physics* 1984; 11:425–447.

Bottomley PA, Hardy CJ, Roemer PB, Weiss RG. Problems and Expediencies in Human 31P Spectroscopy. *NMR in Biomedicine* 1989b; 2(5–6):284–289.

Bronskill MJ, Graham S. NMR Characteristics of Tissue, in *The Physics of MRI: 1992 AAPM Summer School Proceedings*, American Association of Physicists in Medicine,1993, Woodbury, New York.

Brown TR. Nuclear Magnetic Resonance Imaging in Space and Frequency Coordinates. U.S. Patent 4,319,190, 1987.

Bydder M, Larkman DJ, Hajnal JV. Generalized SMASH Imaging. *Magnetic Resonance in Medicine* 2002; 47:160–170.

Callaghan PT. *The Principles of Nuclear Magnetic Resonance*. Oxford University Press, 1993, Oxford.

Carlson JW. An Algorithm for NMR Imaging Reconstruction Based on Multiple RF Receiver Coils. J*ournal of Magnetic Resonance* 1987; 74:376–380.

Carr HY, Purcell EM. Effects of Diffusion on Free Precession in Nuclear Magnetic Resonance Experiments. *Physical Review* 1954; 94(3):630–638.

Chang DC, Woessner DE. Spin–Echo Study of 23Na Relaxation in Skeletal Muscle. Evidence of Sodium Ion Binding Inside a Biological Cell. *Journal of Magnetic Resonance* 1978; 30:185–191.

Cho ZH, Jones J, Singh M. *Foundations in Medical Imaging*. John Wiley & Sons, 1993, New York.

Constantinides CD, Atalar E, McVeigh ER, Signal-to-Noise Measurements in Magnitude Images from NMR Phased Arrays. *Magnetic Resonance in Medicine* 1997; 38:852–857. Erratum in *Magnetic Resonance in Medicine* 2004; 52:219.

Constantinides CD, Westgate CR, O'Dell WG, Zerhouni EA, McVeigh ER. A Phased Array Coil for Human Cardiac Imaging. *Magnetic Resonance in Medicine* 1995; 34:92–98.

Cope FW. NMR Evidence for Complexing of Na+ in Muscle, Kidney and Brain and by Actomyosin. The Relation of Cellular Complexing of Na+ to Water Structure and to Transport Kinetics. *Journal of Applied Physiology* 1967; 50:1353–1375.

Cope FW. Spin–Echo Nuclear Magnetic Resonance Evidence for Complexing of Sodium Ions in Muscle, Brain and Kidney. *Biophysical Journal* 1970; 10:843–858.

Edelstein WA, Foster TH, Schenck JF. The Relative Sensitivity of Surface Coils to Deep Lying Tissues. *Proceedings of the Society of Magnetic Resonance* 1985; 964–965.

Edelstein WA, Glover GH, Hardy CJ, Redington RW. The Intrinsic Signal-to-Noise Ratio in NMR Imaging. *Magnetic Resonance in Medicine* 1986; 3(4):604–618.

Edelstein WA, Hutchison JMS, Johnson G, Redpath T. Spin–Warp NMR Imaging and Applications to Human Whole-Body Imaging. *Physics in Medicine and Biology* 1980; 25(4):751–756.

Ernst R, Bodenhausen G, Wokaun A. *Principles of Nuclear Magnetic Resonance in One and Two Dimensions*. Oxford University Press, 1987, Oxford.

Ernst RR, Anderson WA. Application of Fourier Transform Spectroscopy to Magnetic Resonance. *Review of Scientific Instruments* 1966; 37:93–102.

Freeman R. *A Handbook of Nuclear Magnetic Resonance*. Addison Wesley Longman, 1987, Harlow.

Gadian, DG. *Nuclear Magnetic Resonance and Its Applications to Living Systems*. Clarendon Press, 1982, Oxford.

Gadian DG, Hoult DI, Radda GK, Seely PJ, Chance B, Barlow C. Phosphorus Nuclear Magnetic Resonance Studies on Normoxic and Ischemic Cardiac Tissue. *Proceedings of the National Academy of Sciences of the United States of America* 1976; 73(12):4446–4448.

Gerlach W., Stern O. Der experimentelle Nachweiss der Richtungsquantelung im Magnetfeid. *Zeits. Phys.* 1922; 9: 349–355

Goutsias J. Image Processing Lecture Notes. Department of Electrical Engineering, Johns Hopkins University, 1996, Baltimore.

Grant DM, Harris RK. *Encyclopedia of Nuclear Magnetic Resonance*. Wiley, 1996, New York.

Griswold MA, Jakob PM, Heidemann RM, Nittka M, Jellus V, Wang J, Kiefer B, Haase A. Generalized Autocalibrating Partially Parallel Acquisitions (GRAPPA). *Magnetic Resonance in Medicine* 2002; 47:1202–1210.

Griswold MA, Jakob PM, Nittka M, et al. Partially Parallel Imaging with Localized Sensitivities (PILS). *Magnetic Resonance in Medicine* 2000; 44:602–609.

Gupta RK, Gupta P, Moore RD. NMR Studies of Intracellular Metal Ions in Intact Cells and Tissue. *Annual Review of Biophysics and Bioengineering* 1994; 13:221–246.

Gutowsky HS, McCall DW, Slichter CP. Coupling among Nuclear Magnetic Dipoles in Molecules. *Physical Review* 1951; 84(3):589–590.

Hahn EL. Spin–Echoes. *Physical Review* 1950; 80(4):580–594.

Harpen MD. Noise Correlations in Data Simultaneously Acquired from Multiple Surface Coil Arrays. *Magnetic Resonance in Medicine* 1990; 16:181–191.

Hashemi RH, Bradley WG. *MRI: The Basics*. Lippincott Williams & Wilkins, 1998, Baltimore, MD.

Hayes CE, Edelstein WA, Schenck JF, Mueller OM, Eash M. An Efficient, Highly Homogeneous Radiofrequency Coil for Whole-Body NMR Imaging at 1.5 T. *Journal of Magnetic Resonance* 1985; 63:622–628.

Heidemann RM, Griswold MA, Haase A, et al. VD-AUTO-SMASH Imaging. *Magnetic Resonance in Medicine* 2001; 45:1066–1074.

Hinshaw WS, Andrew ER, Bottomley PA, Holland GN, Moore WS, Worthington BS. An *In Vivo* Study of the Forearm and Hand by the Thin Section NMR Imaging. *British Journal of Radiology* 1979; 52(613):36–43.

Hornak JP. *The Basics of MRI*. 2004. www.cis.rit.edu/htbooks/mri/.

Hoult DI, Busby SJ, Gadian DG, Radda GK, Richards RE, Seely PJ. Observation of Tissue Metabolites Using ^{31}P Nuclear Magnetic Resonance. *Nature* 1974; 252(5481):285–287.

Hoult DI, Lauterbur PC. The Sensitivity of the Zeugmatography Experiment Involving Human Samples. *Journal of Magnetic Resonance* 1979; 34:425–433.

Hoult DI, Richards RE. The signal-to-noise ratio of the nuclear magnetic resonance experiment. *Journal of Magnetic Resonance* 1976; 24(1): 71–85.

Hubbard PS. Nonexponential Nuclear Magnetic Relation by Quadrupole Interactions. *Journal of Chemical Physics* 1970; 53(3):985–987.

Irarrazabal P, Nishimura DG. Fast Three Dimensional Magnetic Resonance Imaging. *Magnetic Resonance in Medicine* 1995; 33:656–662.

Jain AK. *Fundamentals of Digital Image Processing.* Prentice Hall, 1989, Englewood Cliffs, NJ, pp. 11–15.

Jakob PM, Griswold MA, Edelman RR, et al. AUTO-SMASH: A Self-Calibrating Technique for SMASH Imaging; Simultaneous Acquisition of Spatial Harmonics. *MAGMA* 1998; 7:42–54.

Kellman P, Epstein FH, McVeigh ER. Adaptive Sensitivity Encoding Incorporating Temporal Filtering (TSENSE). *Magnetic Resonance in Medicine* 2001a; 45:846–852.

Kellman P, McVeigh ER. Ghost Artifact Cancellation Using Phased Array Processing. *Magnetic Resonance in Medicine* 2001b; 46:335–343.

Kyriacos WE, Panuch LP, Kacher DF, et al. Sensitivity Profiles from an Array of Coils for Encoding and Reconstruction in Parallel (SPACE RIP). *Magnetic Resonance in Medicine* 2000; 44:301–308.

Lamb WE Jr. Internal Diamagnetic Fields. *Physical Review* 1941; 60(11):817–819.

Lauterbur P. Image Formation by Induced Local Interactions: Examples Employing Nuclear Magnetic Resonance. *Nature* 1973; 242(16):190–191.

Lauterbur PC. NMR Zeugmatographic Imaging in Medicine. *Journal of Medical Systems* 1982; 6(6):591–597.

Macovski A. *Medical Imaging Systems.* Prentice Hall, 1983, Englewood Cliffs, NJ.

Mansfield P, Guilfoyle DN, Ordidge RJ, Coupland RE. Measurement of T1 by Echo-Planar Imaging and the Construction of Computer Generated Images. *Physics in Medicine and Biology* 1986; 31(2):113–124.

McVeigh E. *Magnetic Resonance in Medicine: Primer.* Department of Biomedical Engineering, Johns Hopkins University School of Medicine, 1996, Baltimore.

Moon RB, Richards JH. Determination of Intracellular pH by ^{31}P Magnetic Resonance. *Journal of Biological Chemistry* 1973; 248(20):7276–7278.

Morris PG. *Nuclear Magnetic Resonance Imaging in Medicine and Biology.* Oxford University Press, 1986, Oxford.

Noll DC. Multishot Rosette Trajectories for Spectrally Selective MR Imaging. *IEEE Transactions of Medical Imaging* 1997; 16(4):372–377.

Noll DC. Recent Advances in MRI. Presentation, Michigan University.

Ogawa S, Menon RS, Tank DW, Kim SG, Merkle H, Ellermann JM, Ugurbil K. Functional Brain Mapping by Blood Oxygenation Level-Dependent Contrast Magnetic Resonance Imaging. *Biophysical Journal* 1993; 64(3):803–812.

Overhauser AW. Polarization of Nuclei in Metals. *Physical Review* 1953; 92(2):411–415.

Pauly J, Nishimura D, Macovski A. A *k*-space analysis of Small-Tip-Angle Excitation. *Journal of Magnetic Resonance* 2001; 213(2):544–557.

Proctor WG, Yu FC. The Dependence of a Nuclear Magnetic Resonance Frequency upon Chemical Compound. *Physical Review* 1950; 77(5):717.

Pruessmann KP, Weiger M, Scheidegger MB, Boesinger P. SENSE: Sensitivity Encoding for Fast MRI. *Magnetic Resonance in Medicine* 1999; 42:952–962.

Purcell E, Torrey HC, Pound RV. Resonance Absorption by Nuclear Magnetic Moments in a Solid. *Physical Review* 1946; 69:37–38.

Ramo S, Whinnery T, Van Duzer T. *Fields and Waves in Communication Electronics.* John Wiley & Sons, 1984, New York.

Ramsey NF. A Molecular Beam Resonance Method with Separated Oscillating Fields. *Physical Review* 1950; 78(6):695–699.

Roemer PB, Edelstein WA, Hayes CE, Souza SP, Mueller OM. The NRM Phased Array. *Magnetic Resonance in Medicine* 1990; 16:192–225.

Schmitt M, Potthast A, Wald L, Sosnovick D. Radiofrequency Coil Innovation in Cardiovascular MRI. Technology and Trends, *Magnetom Flash* 2/2007. www.siemens.com/magnetom-world.

Shellock FG. *Magnetic Resonance Procedures: Health Effects and Safety.* CRC Press, 2001, Boca Raton, FL.

Shporer M, Civan MM. Nuclear Magnetic Resonance of Sodium-23 Linoleate Water. *Biophysical Journal* 1972; 12:114–122.

Shporer M, Civan MM. Effects of Temperature and Field Strength on the NMR Relaxation Times of 23Na in Frog Striated Muscle. *Biochemical Biophysical Acta* 1974; 354:291.

Slichter CP. *Principles of Magnetic Resonance.* Springer, 1990, Berlin.

Sodickson DK, Manning WJ. Simultaneous Acquisition of Spatial Harmonics (SMASH): Fast Imaging with Radiofrequency Coil Arrays. *Magnetic Resonance in Medicine* 1997; 38:591–603.

Sprawls P, Bronskill MJ. *The Physics of MRI: 1992 AAPM Summer School Proceedings.* American Association of Physicists in Medicine, 1993, Woodbury, NY.

Springer CS. Measurement of Metal Cation Compartmentalization in Tissue by High-Resolution Metal Cation NMR. *Annual Review of Biophysics and Biophysical Chemistry* 1987; 16:375–399.

Stark D, Bradley W. *Magnetic Resonance Imaging.* Mosby, 1999, St. Louis, MO.

Thomas SR. Magnets and Gradients Coils: Types and Characteristics, in *The Physics of MRI: 1992 AAPM Summer School Proceedings*, American Association of Physicists in Medicine, Woodbury, New York, 1993.

Wang J, Kluge T, Nittka M, et al. Parallel Acquisition Techniques with Modified SENSE Reconstruction mSENSE. In *Proceedings of the First Wurzburg Workshop on Parallel Imaging Basics and Clinical Applications*, Wurzburg, Germany, 2001, p. 89.

Index